Science Was Born of Christianity:
The Teaching of Father Stanley L. Jaki

Science Was Born of Christianity

The Teaching of
Father Stanley L. Jaki

STACY A. TRASANCOS

The Habitation of Chimham Publishing Company

© 2014 by The Habitation of Chimham Publishing
http://www/chimhampublish@aol.com
chimhampublish@aol.com

Paperback
All Rights Reserved

Library of Congress Cataloging-in-Publication-Data

Trasancos, Stacy
Science Was Born of Christianity

ISBN: -13-978-0-9899696-1-1
Library of Congress Control Number 2014940470

1-Science – history 2-Christianity 3-Philosophy

with a review written for inclusion in a magazine, newspaper, website, or broadcast.

To my children.

Sancta Maria, Mater Dei,
ora pro nobis peccatoribus,
nunc, et in hora mortis nostrae.
Amen.

Acknowledgements

The research for this book was compiled for a master's thesis submitted in partial fulfillment of the requirements for the degree of Master of Arts in Theology from Holy Apostles College and Seminary in Cromwell, Connecticut. I would like to thank my research advisers, Dr. Alan Roy Vincelette, Ph.D., who is an Associate Professor at St. John's Seminary in Camarillo, California, and Dr. Donald DeMarco Ph.D., who is a Senior Fellow of Human Life International and Professor Emeritus at St. Jerome's University in Waterloo, Ontario. Both professors provided guidance in the compilation and presentation, softened a claim here and added another section there, all while allowing me the freedom to write with my own voice.

I am grateful to Dr. Sebastian Mahfood, Associate Professor of Interdisciplinary Studies, Vice-President of Administration, Director of Distance Learning, and Director of Assessment, for the organization and execution of the distance learning program at Holy Apostles. The program has matured in the three years I attended (from the comfort of my home and with babies in my lap), and he is undoubtedly committed to furthering the excellence of the college.

I thank my husband, Jose Trasancos, for his love, support, and patience during the time I devote to studies. We both share an admiration for Fr. Jaki, this priest and physicist we never met. My husband indulges the continuance of my education and the building of my personal library, a patron if ever there was one.

I thank the people who read and edited this book. Mr. Antonio Giovanni Colombo read several drafts to provide feedback on consistency with Fr. Jaki's work since he was a friend of Fr. Jaki's and manages the collection of Jaki's work at Real View Books. Mr. Colombo also provided help with the accuracy of the sources and references.

I thank my friend, prayer partner, and fellow scientist, Dr. Walter Bruning, for his continual straightforward feedback on the tone of my writing and for his encouragement in my faith and intellectual pursuits.

I thank Dr. Jeff McLeod, an Adjunct Associate Professor of Psychology at St. Mary's University of Minnesota and faculty member at the University of St. Thomas in St. Paul, Minnesota, where he teaches at the St. Paul Seminary in the Archbishop Harry J. Flynn Catechetical Institute. He encouraged me to keep writing and to view the questions of science and religion in a broader context. I do not think Dr. McLeod ever had a word of criticism, but his feedback demonstrated that he always understood what I was thinking, which is wonderful feedback for a writer.

I thank Associate Professor Emeritus of Theology at Christendom College, Dr. John Janaro, for graciously offering to read my draft with an editorial eye and for discussing the meaning of science according to Fr. Jaki. Dr. Janaro knew Fr. Jaki and worked with him to publish his work during his lifetime.

I thank Mr. John Darrouzet, a Hollywood screenwriter and accomplished lawyer, who gave me enthusiastic feedback on this book from a layman's perspective. He understands my passion for communication in plain language.

I also thank Mrs. Cynthia Trainque, a friend and fellow theology student, for her guidance on writing the introduction.

Royalties from this book go to a friend, a United States military veteran and single mother, beginning on its publishing date, December 6, 2013, the Feast Day of St. Nicholas, and extending for as long as she accepts the gift.

Finally, I thank you, the reader, for giving me a chance to explain why I admire the work of the shepherd Fr. Stanley L. Jaki.

Contents

Foreword

Foreword

I warmly recommend this book written by Stacy Trasancos on the Christian basis and inspiration for the birth of modern science according to Fr. Stanley Jaki OSB, the great philosopher of science and theologian, member of the Pontifical Academy of Sciences and Templeton prize-winner. The author has made a great effort and has succeeded in making Fr. Jaki's ideas available to a larger public. Her goal is to inform more people about what Fr. Jaki actually meant by this claim that Christianity stimulated the birth of science because it is significant for people (young adults especially) to understand that faith and science are not opposed, and are indeed complementary Also Stacy realizes that it is necessary to set the record straight, because so many people, even Catholics, misrepresent what Fr. Jaki actually concluded about the birth of science.

In a world swirling with relativist and materialist notions concerning the origins of the cosmos and of the human person, Stanley Jaki has offered scientists, philosophers and seekers alike a way out of this morass. He explains how the idea of the beginning of the cosmos, which is so much part of Christian tradition, stands in sharp contrast to the scene outside of Christianity where many world religions and world-pictures had great difficulty in maintaining that the world actually began. Even for many people today, the world is eternal in the sense that it simply is. The world was often regarded as eternal in seven principal ancient cultures: Chinese, Hindu, Meso-American, Egyptian, Babylonian, Greek and Arabic. From the cosmic imprisonment represented by all these world pictures, Christianity was to bring liberation. All ancient cultures held a cyclic view of the world, and this was one of the beliefs that hindered the development of science. This cyclic pessimism was decisively broken by the belief in the unique Incarnation of Christ;

thereafter time and history were seen as linear, with a beginning and an end.

If science suffered only stillbirths in ancient cultures, this implies that it arrived more recently at its unique viable birth. The beginning of science as a fully-fledged enterprise can be said to have taken place in relation to two important definitions of the Magisterium of the Church. The first was the definition at the Fourth Lateran Council, in the year 1215, that the universe was created out of nothing at the beginning of time. The second magisterial statement was at local level, enunciated by Bishop Stephen Tempier of Paris who, on 7 March 1277 condemned 219 Aristotelian propositions, so outlawing the deterministic and necessitarian views of the creation. These statements of the teaching authority of the Church expressed an atmosphere in which faith in God the Creator had penetrated the medieval culture and given rise to philosophical consequences. The cosmos was seen as contingent in its existence and thus dependent on a divine choice which called it into being; the universe is also contingent in its nature and so God was free to create this particular form of world among an infinity of other possibilities. Thus the cosmos cannot be a necessary form of existence and so has to be approached by a posteriori investigation. The universe is also rational and so a coherent discourse can be made about it. Thus the contingency and rationality of the cosmos are like two pillars supporting the Christian vision of the cosmos.

In the Middle Ages, ideas about the created universe had developed which were greatly conducive to scientific enterprise. The philosophical vision of the Christian Middle Ages perceived the cosmos as demythologized, free from the capricious whims of pantheistic voluntarism reified in pagan deities. This world vision included the idea that the cosmos is good, and therefore attractive to study. Also the universe was considered to be single entity with inner coherence and order, and not a gigantic animal which would behave in an arbitrary fashion, as was often believed in antiquity. The unity of the universe offers a challenge to investigators to search for the connections in nature and make them explicit. Further, the cosmos was seen to be rational and consistent, so that what was investigated one day would also hold true the next. This encouraged repetition and verification of experiments. The world picture also involved the tenet that cosmic order is accessible to the human mind, and needs to be investigated experimentally, not just by pure thought. The world was considered to be endowed with its own laws which could be tested and verified; it was not magical or divine. In addition to these ideas, medieval Christendom also was imbued with the concept that it was worthwhile to share knowledge for the common good. Finally the cosmos was seen as beautiful, and therefore investigation of it gave a participation in such beauty which elevated the mind and heart of the believing scientist to the Creator.

In short there is truly a mine of wisdom in these pages, and Stacy Trasancos has not only faithfully transmitted the teaching of Fr Stanley Jaki in a form which may be readily digested by today's public, but she has also carried out a service to the intellectual peripheries of society, which are hungry for the truth of the Gospel, as Pope Francis would put it.[1]

Rev. Dr. Paul Haffner,
President, Stanley Jaki Foundation
Invited Lecturer Pontifical Gregorian University
Associate Professor, Duquesne University Italian Campus
Rome, 19 March 2014
Solemnity of St. Joseph, Patron of the Universal Church

---

[1] Pope Francis, *Evangelii Gaudium*, 1: "Each Christian and every community must discern the path that the Lord points out, but all of us are asked to obey his call to go forth from our own comfort zone in order to reach all the 'peripheries' in need of the light of the Gospel."

Introduction

Introduction

It is a daunting task to analyze the work of a scholar and formidable researcher as profound as the late Father Stanley L. Jaki. He was a Benedictine priest, theologian, and physicist who was awarded the 1987 Templeton Prize for being a leading thinker in areas at the boundary of science and theology. One of Jaki's most useful teachings was the application of Gödel's incompleteness theorem as a "cudgel against those scientists who try to shore up their materialism with their 'final' cosmologies."[1] Such theories in physics must be heavily mathematical, and thus, per Gödel's theorem, they cannot have in themselves proof of their own consistency. Neither can Creation be fully grasped by physics, since scientists cannot go outside the cosmos to measure it. It is akin to a child trying to prove a final theory of how his house works when he is incapable of toddling outside its doors.

Likewise, an effort to represent Jaki's teaching when one is not Jaki, but instead a student attempting to understand and convey what is gained from his findings, also has a feeling of incompleteness to it, an inability to wrap arms around the whole of it and comprehend it adequately, and thus, a hesitation to go from reader and explorer of his work to writer and analyzer reflecting on it, such a sentiment Jaki would perhaps appreciate since he was also fond of Blessed John Henry Newman's saying, "Nothing would be done at all if a man waited until he could do it so well that no one would find fault with it."[2] The same is no less true, obviously, for a woman.

---

[1] Stanley L. Jaki, *Science and Religion: A Primer* (Port Huron, MI: Real View Books, 2004), 12; Stanley L. Jaki, *The Drama of Quantities* (Port Huron, MI: Real View Books, 200), 25-50.

[2] Stanley L. Jaki, *A Mind's Matter: An Intellectual Autobiography* (Cambridge, UK: William B. Eerdmans Publishing Company, 2002), 103.

The first time I read Jaki's book *The Savior of Science*, I knew I was reading something significant. It was during my tenth pregnancy and first graduate theology course. Having left a position as an industrial research scientist to become a full-time homemaker, my interests as a Catholic convert turned toward the intersections of science and religion. I recognized the scientist in Jaki right away; how it would take some years for me to grasp the theological and historical implications of his claim about the birth of science. I not only tried to understand it with the mind of a scientist and a theologian, but also with the heart of a mother; that is how the "birth" and "stillbirth" analogies put forth by Jaki most vividly took root in my mind.

I understood from my first reading of Jaki's book that he was arguing that science was intimately connected with Christianity, but when I tried to articulate how "science was born of Christianity" after being "stillborn" in other cultures, I was unable to satisfactorily grasp the whole idea and explain it in a substantial way. Jaki not only argued that Christianity in the Middle Ages contributed to the rise of modern science and that the Church was a patron of science, but he went even further; he argued that there *had to come a birth*, the birth of the only begotten Son of the Father as a man, to allow science to have its first viable birth. To argue that is not a trivial matter.

The person who wishes to understand this claim needs to first understand that it is more than a claim that Christianity informed man that he was made in the image of God and that the world was ordered. To claim that is to claim that science is a human endeavor that began with the beginning of human existence. This historical research is more specific. It is about how faith in divine revelation produced the breakthrough in an understanding of the universe that caused a departure from ancient worldviews of an eternally cycling universe, and led to the breakthrough that was necessary for the Scientific Revolution to occur. This departure, this *breakthrough*—this birth—was not based on observation or experiment but on

16

faith in the Christian Creed. The purpose of this short book, therefore, is to explain Jaki's assertion that "science was born of Christianity," having been "stillborn" in other ancient cultures. If by the end of the book you can at least articulate his position, whether you agree with it or not, then I have been successful.

Jaki's work has been misunderstood and often misrepresented. It is at risk for being marginalized or even dismissed altogether. His conclusions are, in my opinion, increasingly relevant as scientific advances are made. It is also important to know why Jaki did this work. Jaki furthered this particular assertion not so that Christians could triumphantly prove the superiority of Christian thought, an often made dismissive assumption, but because science of the future depends on an understanding of what science is and how it came to be. So much of modern humanity is affected by science and can even be destroyed by science. If science is to progress and to benefit the human race, it must return to the protective mantle that Christian thought provided it when it was born.

Why does this matter? It matters because science today is returning to the pantheistic, pagan, or atheistic thought of ancient times—times when science did not thrive as a self-sustaining enterprise that discovered physical laws and systems of laws, but instead viewed the world as unpredictable, unknowable, and magical. This view will be explained as it is the essence of the argument. To understand this claim is to understand why the Catholic Church has a legitimate right and authority to challenge scientific conclusions which directly contradict divinely revealed dogma. It is also to understand that to approach science in this way is to approach it the same way scholars did when modern science was born, thrived, and matured. It is to understand how to sort through the issues in the frontiers of science. That is the reason this claim, when presented correctly, ought to be of interest to people whether Catholic or not. Furthermore, this claim well may prove to be a useful tool for evangelization if presented carefully.

Who Was Fr. Jaki?

Who Was Fr. Jaki?

Having authored over fifty books on the history and philosophy of science and natural theology, Jaki's legacy has yet to be fully realized because it is no small effort to read his work. He knew so much, researched so thoroughly, and was unafraid to make bold statements to defend his findings; it is easy for a newcomer to Jaki's thought to fail to appreciate what it can teach. A dedicated student needs to study several books before the overall message aggregates and penetrates, but once that effort is made, it becomes clearer how united and pervasive his thought was—and actually how simple the whole of it is.

Jaki was a priest, a theologian, a physicist, a historian, and a philosopher, possessing a vastness of knowledge that gave him a breadth of insight that a scholar acting as any one of those titles alone cannot achieve. He didn't study history, for instance, as just facts; he studied history while also considering theology, culture, technology, prose, and artwork so that he could find the origins for certain cultural psychologies that affected decisions. His approach to history has been called both evolutionary and revolutionary.[1] Here is a brief review of some of the relevant works.

In 1966, after earning doctoral degrees in theology and physics, Jaki published a six-hundred-page volume detailing the history of physics using original sources at the time available mostly in the Firestone Library of Princeton University.[2] He titled it *The Relevance of Physics*, and it was published by Chicago University Press. It was reprinted in 1970 and in 1992. The content of *Relevance of Physics* was

---

[1] He described it this way himself in the 1986 preface to *Science and Creation: From Eternal Cycles to an Oscillating Universe* (Edinburgh: Scottish Academic Press, Ltd, 1986), vi.

[2] Jaki, *A Mind's Matter*, 33.

completed in 1969 by *Brain, Mind, and Computers*, for which Jaki was given the Lecomte de Nouy prize in 1970. After that he wrote about subjects related to the history of astronomy in a series of articles, and in a few books, namely *The Paradox of Olbers' Paradox* (1969), *The Milky Way: An Elusive Road for Science* (1972) and *Planets and Planetarians: A History of Theories of the Origin of Planetary Systems* (1978).

In 1969, he wrote a preface to the English translation of a work of Pierre Duhem, *To Save the Phenomena: An Essay on the Idea of Physical Theory from Plato to Galileo*. This is the first essay he devoted to Duhem, whom he quoted in *The Relevance of Physics*. After conducting more research of Pierre Duhem's work and the origins of science, Jaki published *Science and Creation: From Eternal Cycles to an Oscillating Universe* in 1974, in which he carefully considered the fortunes and misfortunes of science in all the main ancient cultures: India, China, pre-Columbian America, Egypt, Mesopotamia, and Greece, as well as the birth of science in Christian Europe. In 1978, he published *The Road of Science and the Ways to God: The Gifford Lectures 1975 and 1976*, a history of science, and later, *The Origin of Science and the Science of its Origins*, the enlarged text of a series of Fremantle lectures given at Balliol College, Oxford the previous year. In 1984, he published *Uneasy Genius: The Life and Work of Pierre Duhem* after intensive research of his "kindred mind" who died eight years before Jaki was born.[3] Jaki literally traced Duhem's footsteps during his lifetime in Bordeaux and credited Duhem with the historical research of original sources that provided the evidence that the birth of science is owed to the mindset of the Christian Middle Ages.

The historical research culminated in a book first published in 1988 which contains the lectures that formed the basis of discussions at the Third Annual Wethersfield Institute Conference in New York and a lecture given at Columbia University, both in 1987. To his admitted "sheer

---

[3] Jaki, *A Mind's Matter*, 69.

delight," Jaki titled the book *The Savior of Science*.[4] Since then he gave numerous lectures on the subject to summarize it, and those have been published as well.[5] This brief timeline of publication is far from complete, but those are the books most related to the content of this book. Together they comprise an enormous collection of research and insight. The primary sources used are *Science and Creation, The Savior of Science,* and his autobiography, *A Mind's Matter: An Intellectual Autobiography,* along with other works and essays from Jaki's later years.

This is also not the first book to be written about Jaki and his teaching. In 1991, a priest and physicist, Father Paul Haffner, published a doctoral thesis to portray Jaki's contributions to the understanding of the relationship between Christianity and modern science.[6] His book, *Creation and Scientific Creativity: A Study in the Thought of S. L. Jaki,* is the first systematic treatment of the first thirty years of Jaki's studies, a topical and synthetic presentation of Jaki's doctrine. It is valuable to anyone who wants a survey of Jaki's research and perceptions prior to the 1990's, particularly the theological aspects.[7] Haffner was fortunate in that he was able to consult with Jaki during his doctoral work, and he generously included a chapter on Jaki's life and career. Haffner also compiled, with Jaki's aid, the first full bibliography of Jaki's publications. Then in 2009, to commemorate Jaki's death, Haffner published an expanded and revised edition of his book, which included a complete sixty-page bibliography of Jaki's entire publications, theses, lectures, essays, and books. Haffner also added a list of

---

[4] Jaki, *A Mind's Matter*, 49.

[5] Most are published by Real View Books, and are available on that website.

[6] Paul Haffner, *Creation and Scientific Creativity: A Study in the Thought of S. L. Jaki*, 2nd Edition (Herefordshire: Gracewing, 2009).

[7] Haffner, 7.

reviews of Jaki's work organized by general reviews and particular reviews related to his major books.[8]

Since the 1991 publication of Haffner's thesis, Jaki published at least twenty more works, continually refining his thought and furthering it into philosophy and social issues. In his 2002 autobiography, *A Mind's Matter*, published in his late seventies, Jaki used the simile of crystal growth to describe his intellectual maturation, a sharpening of the mind as if the edges slowly became more rigid so that the transparency became lucid. Jaki exhaustively, almost superhumanly, referenced his early works with original sources, something he insisted on and travelled great distances and invested much time to acquire if necessary, although he accessed a great deal of original works in the libraries at Princeton. His earlier volumes are large and intense with characteristic Jakian meticulousness. They form the foundation of the more crystallized ideas in his later works.

Jaki's later monographs, essays, and lectures by contrast, have a reflective tone to them and read more as syntheses that cultivate his conclusions. They read quite easily but assume the reader has some background or knowledge of the material. Jaki's work could probably best be approached in the following manner. Reading the earlier volumes is akin to witnessing how a monument is built stone by stone, the only patient endeavor that can give an onlooker the fullest appreciation for the quality of the construction. Reading the later works is more like being given a guided tour of the monument by the builder himself.

Even though almost all historians of science now support the contributions of the Roman Catholic Church during and before the Middle Ages, even Catholic historians of science are reluctant to embrace Jaki's theological argument that the Christian worldview was responsible for the birth of science. Undoubtedly Jaki's work contributed to the easing of the

---

[8] Haffner, 311-320.

myth of conflict between science and religion that was very popular from the time of the Enlightenment to the decades in which Jaki worked. Therefore, it is disappointing that he is insufficiently credited as a leading proponent of that cause and that his argument about the theological origins of science is not more widely known, a disappointment this book seeks to rectify.

The chapters are organized to define and defend the title as it goes, "Science Was Born of Christianity," and are organized into an outline for easy reference. The chapter following this introduction is titled "Science," and it explains how Jaki clarified the definition of science in contrast with religion, and why he insisted that there can never be a conflict between the two. Those definitions must be understood before the rest of the argument can be grasped, for it is the lens through which Jaki searches history.

The next chapter, "Was Born," is the longest. It is a four-part review of Jaki's historical findings of the "stillbirths" of science in other ancient cultures and the "birth" of science in the Middle Ages of the Christian West. This information is extensively provided in *Science and Creation* and is necessary background information. Since covering such a vast history is, of course, impossible in a brief overview, the scientific successes of each culture are described followed by the reasons for the stillbirths of science within each one, as Jaki saw it. The second section covers the Old Testament worldview and how that biblical view purified a truly scientific worldview, a little appreciated point that is actually rather obvious when understood. The third section of this chapter addresses the scientific attitudes of the early Church Fathers, the point being to show the continuity from the biblical cultures into the Middle Ages. The last section is a lengthy survey of medieval Christian scholars demonstrating the breakthrough in scientific thought that culminated because of the Christian Creed, a breakthrough that can be taken for the birth of modern science.

The next chapter, "Of Christianity," goes into more detail about Jaki's rationale that Christian belief was the cultural matrix, or "womb," from which science was born. This is the theological aspect of Jaki's argument, and it is the aspect most ignored and misunderstood. A careful reading of it should tie the previous two chapters together to give the full weight of Jaki's accomplishment. In the next chapter, the opinions and misgivings of Jaki's critics are addressed. Anyone arguing that science was born of Christianity must be aware that, if done carelessly, it may offend and be received as arrogant, exclusionary, or chauvinistic. There is an understandable risk that such an argument will sound condescending to other religions while self-congratulating Christianity.

The final chapter is my commentary on the significance of Jaki's research and insight and some thoughts for furthering this work with an ecumenical frame of mind.

Chapter 1 – "Science"

Chapter 1 – "Science"

"And it is always with measurement that the buck stops with science."[1] That line is basically Jaki's definition of science. The more technical definition is this: *Exact* science is the *quantitative* study of the quantitative aspects of *objects* in motion. The words "exact" and "quantitative" and "object" should be noted, and the fact that the very word "definition" means to put boundaries around something. This chapter will explain why Jaki insisted on this definition and why this definition matters in its relation to religion.

It is an unusually short definition, but it takes a much longer explanation because the word "science" has meant so many things. Today it has especially become confused by the advance of the phenomenon known as "Scientism," the belief that science can solve far more problems of humanity than it actually can. Thus, there is a prevalent mistake in modern culture to overstep the limits of science.

Jaki not only clarified the definition of science, he applied it in his work. Where some historians have tried to understand the history of science by trying to understand what science meant in different times and cultures, Jaki approached the question the other way. He first defined that human endeavor of investigation and understanding strictly, and then he searched through history to discover where it was born and where it was not and, most importantly, *why* it was born and why it was not. Jaki's purpose was not just to tell the story of science; it was to view that history through a scientific and theological lens so that the present condition and future of science could be better understood. If you are not aware of this short definition upon reading Jaki's works, it can be puzzling.

---

[1] Stanley L. Jaki, *A Late Awakening and Other Essays* (Port Huron, MI: Real View Books, 2004), 68.

As a physicist, Jaki consistently referred to science as "exact science." He used that term in his doctoral research conducted from 1956 to 1958 and still in his 2004 essay, *Science and Religion: A Primer.*[2] In the primer, which was inspired during a Chestertonian conference held in Minneapolis of 2004, he began the discussion with, "By science exact science is meant throughout this booklet."[3] He ended the epilogue of the primer with:

> Equations of numbers are practically everything in science, very little in philosophy, and nothing in theology. It is therefore a huge mistake to take trendy philosophies of science, let alone some theological flights of fancy, for science. Numbers alone make science.[4]

He covered this distinction more thoroughly in a 2003 compilation of essays, *Numbers Decide*, a 2004 book, *Questions on Science and Religion*, and again in his 2006 collection of essays, *A Late Awakening and Other Essays*. He discussed this distinction in *The Savior of Science*, but without realizing how pervasively Jaki used this insistence that science be exact, quantitative, and about objects in motion, the significance may go unnoticed. You need that background before attempting to understand Jaki's argument that *science* was born of Christianity.

It is obvious today that science is about the quantitative aspects of objects in motion. On the grand scale, the ability to travel in space has been developed by science. On the minute scale, particle accelerators to detect the motion of subatomic particles have been developed by science. Modern laboratories are designed to trap, manipulate, or measure moving objects. The equipment aims radiation or electrons at

---

[2] Haffner 13.
[3] Jaki, *A Mind's Matter*, 24; Jaki, *Science and Religion*, 4.
[4] Jaki, *Science and Religion*, 31.

objects, moves gas particles onto surfaces, captures photo-induced fluorescence within molecules, measures the mass of atoms or molecules produced upon decomposition, or charts flow rates through pipes or drums, to name a few examples. In technology the quantitative movement of objects has delivered "ever more stunning marvels," from the harnessing of the flow of electrons in metal wires to the the detection of waves triggered by their acceleration in antennas, or "jumps" between "holes" in semiconductor materials.[5] All of these marvels imply a continued reliance on Newton's three laws of motion: 1) by the law of inertia, a body remains at rest or moving uniformly in a motion unless acted on by an external force; 2) the acceleration (a) of a body is proportional to the force (F) acting on it and inversely proportional to its mass (F=ma); and 3) to every action there is an equal and opposite reaction.[6] That is to say, all these marvels of science imply a continued reliance on "the *quantitative* study of the quantitative aspects of *objects* in motion," thus Jaki's definition. It was based on what is generally understood to be modern science, although more concise than other definitions.

Jaki's definition was also based on ancient physics. Numbers are the only specifically exact notions, among all other notions, that the human mind is capable of forming. Aristotle recognized that numbers stand apart some 2,300 years ago.[7] He recognized that there are *quantities*, and there are *qualities*. Quantities are numerical, qualities are not. Jaki drew this distinction of quantities versus qualities from a dictum in Aristotle's *Categories*, and it hinged on three little

---

[5] Stanley L. Jaki, *The Savior of Science* (Grand Rapids, MI: William B. Eerdmans Publishing Company, 2000), 50.

[6] Jaki, *Savior of Science*, 50; Isaac Newton, *Principia, A New Translation* by I. B. Cohen and A. Whitman (Berkeley, CA: University of California Press, 1999), 416-417.

[7] Jaki, *Science and Religion*, 4.

words, "more or less."[8] In *Categories*, Aristotle enumerated the ten categories that can describe every object, using the phrase "more or less" repeatedly.[9] He wrote, "There is nothing that forms the contrary of 'two cubits long' or of 'three cubits long', or of 'ten', or of any such term. A man may contend that 'much' is the contrary of 'little', or 'great' of 'small', but of definite quantitative terms no contrary exists."[10]

Aristotle went on to explain that if something is white, it can be "more or less" white.[11] If something is beautiful, it can be "more or less" beautiful than another object.[12] Habits can be "more or less" permanent. In the grasping of honor, men may be "more or less" brave or practice justice and self-mastery "more or less."[13] But "more or less" cannot be predicated on quantities; quantities are absolute. There is no more or less to the number "1," for instance, or to any other number or numerical fraction of a number.[14] Quantities–and quantities alone–are exact. "Quantity does not, it appears, admit of variation of degree," Jaki quoted from Aristotle.[15] About words belonging to all of the other nine categories, the actions, the qualities, the states of existence, including time, place, and affection, one can say "more or less." One can be more or less running. One can be more or less kind. One can be more or less a child. One can be more or less in the

---

[8] Aristotle, *Categories*, Part 5. For Jaki's explanation see Jaki, *Science and Religion*, 28; Jaki, *A Late Awakening*, 63. See also Stanley L. Jaki, *Numbers Decide and Other Essays* (Pinckney MI: Real View Books, 2003), 191; Stanley L. Jaki, *Questions on Science and Religion* (Pinckney MI: Real View Books, 2004), 183; and Stanley L. Jaki, *The Drama of Quantities* (Port Huron, MI: Real View Books, 2005).

[9] Aristotle, *Categories*, Part 5.

[10] Aristotle, *Categories*, Part 5.

[11] Aristotle, *Categories*, Part 5.

[12] Aristotle, *Categories*, Part 5.

[13] Aristotle, *Categories*, Part 8.

[14] Aristotle, *Categories*, Part 5.

[15] Aristotle, *Categories*, Part 6.

morning. One can be more or less at home. One can be more or less in love. Yet for numbers, "more or less" can never be said. The number "one" cannot be more or less one.

A Thomistic scholar will recognize that this definition of "science" differs from the classical definition in the medieval universities, and, indeed, it is not how St. Thomas Aquinas used the word in the beginning of the *Summa Theologiæ*, where he stated explicitly that sacred doctrine (theology) is a science, *scientia*, because Sacred Scripture "considers things precisely under the formality of being divinely revealed."[16] The term *scientia* means "knowledge," and modern science of course did not get its meaning until much after Aquinas. Theology, according to St. Thomas, is a science, the highest science with God as its object; it is founded on the Wisdom of God. To compare Jaki's definition with St. Thomas' classical definition is too lengthy a topic to explore in the present work, but it is sufficient to point out that Jaki's use of the term "exact science" does not conflict with St. Thomas' ordering of theology and other sciences.[17] Jaki addressed the comparison in more detail in the two last essays in his 2006 publication *A Late Awakening*, "A Non-Thomist Thomism" and "Thomas and the Universe."[18]

Modern science has revealed aspects of the quantification of the natural world that were unknown in St. Thomas' time. There was not a word for what today is understood as Newtonian science, but that does not mean that such a word should not be found. To update Thomism is not to oppose Thomism, but rather to appreciate the thought of Aquinas so much as to want to keep it updated as new insights are gained. In the modern world, there is a vast confusion about science, and that confusion stems from a departure from

---

[16] *ST*, I, q. 1, a. 2, *Respondeo dicendum*, Latin-English Edition, vol. 1, NovAntiqua.

[17] *ST*, I, q. 1, a. 5.

[18] Jaki, *A Late Awakening*, 217-256.

quantities. Such a departure from quantities as intrinsic to science allows for erroneous philosophies to be grafted onto science, and in turn to be grafted onto Thomistic thought to conjure up notions of Neothomism and a transcendental Thomism—something Jaki termed (pejoratively) Aquikantism.[19] That is a subject for another analysis, so back to Aristotle.

Of all Aristotle's other nine categories that describe objects, quantities stand in "splendid isolation" from other categories of words. That is the point. The overwhelming majority of propositions the human mind must address to understand reality—and here is the brilliance of this axiom—have nothing to do with numbers.[20] In antiquity, nature was described mostly qualitatively, but in modernity, quantitative measurements are required to define physical laws. This change to quantities marks the Scientific Revolution that transformed the world. In Aristotelian physics, "more or less" was enough because Aristotle's physics noted details qualitatively of substances in the natural world, but Newtonian physics changed that. Newtonian physics addresses every proposition aiming for *quantitative* exactness, and "more or less" is no longer appropriate. The Scientific Revolution was a change to recognize the significance of exact quantities.

Since exact science derives its exactness from measurements, the application of quantities, it is *limited to matter*, and since science is limited to matter, it has no philosophical or religious implications whatsoever. "Herein," Jaki wrote, "lies the crux of all talks about science and religion, and even about science and any non-scientific field of inquiry, generally known as the humanities."[21]

---

[19] Jaki, *A Mind's Matter*, 100.

[20] Jaki, *Science and Religion*, 4-8; Jaki, *Numbers Decide*, 191-196.

[21] Jaki, *Science and Religion*, 4.

At this point, in trying to assimilate this definition, you are probably asking, "But what about chemistry, biology, and the other fields that are scientific, such as medicine, political science, and social science?" There absolutely needs to be an answer for that. Jaki sometimes further clarified that by "exact science" he meant physics, and because he was a physicist, it is understandable that he started there. He understood that physics was the purest quantitative science: "Physics, to restate, with some emphasis, a definition I give in various publications 'is the *quantitative* study of the *quantitative* aspects of things in motion.'"[22] That definition is also his definition of exact science.

However, Jaki did not stop with physics; he used the term "exact science" to emphasize that which is "exact" knowledge from that which is not exact, which he called "reasoned discourse." Since mathematics does not measure, it is a form of logic. Physics and astronomy are the most exact sciences since they are the quantitative study of the quantitative aspects of objects in motion. Biochemistry and biophysics are becoming exact sciences because they are becoming fields that measure objects in motion. Chemistry is an exact science that deals with averages of interactions between many objects. Evolutionary biology is an exact science only so far as it measures quantities and is about mechanism, but when extended as an explanation beyond that, it is reasoned discourse, and in the case of materialistic Darwinism, unreasonable discourse. Neurology is an exact science only when it measures the brain or nervous system, but it cannot measure the soul. A characteristic of exact science is that, being based on measurements, it allows exact predictions. Psychology, sociology, and political science are minimally exact sciences where–and only where–they concern

---

[22] Jaki, *Questions*, 11. In Jaki, *A Mind's Matter*, 178-179 he clarified that "objects" is the better word instead of "things."

quantified data collected by measurement.[23] For the most part, those disciplines are matters for philosophical consideration. Jaki went so far as to say that theology should not be called a science because it rests on considerations that are not quantitative.[24]

Jaki pointed out, however, that even "the understanding of quantities rests on non-quantitative propositions."[25] Newton's laws of motion explain quantitative data, but even in the strictest sense, universally non-numerical words are needed to give any meaning to numerical quantities. Therefore, reasoned discourse stands at the beginning and at the end of quantities, embedding science within it. Exact science depends on reasoned discourse.

Jaki did not insist on the distinction between "exact science" and "reasoned discourse" to diminish science, but to emphasize the *power* of reasoned discourse. As a brief but necessary side note, there is a theological underpinning to this emphasis related to the doctrine of *imago Dei*, that man is made in the image of God. Scripture revealed that the Holy Trinity is an ordered procession. The Father generates the Son as a divine internal act of the intellect (which is why the Son is also called the Word or the *Logos*), and the Father and Son together as one substance spirate (from *spirāre*, breathe forth) the Holy Spirit as a divine internal act of the will. Thus, humans created with a rational soul also have intellect and

---

[23] A good summary of this definition is given in the booklet *Science and Religion*, 4-17. Jaki also addresses it in many other places, for instance *Questions*, Chapter 14; *Drama of Quantities*, Foreword; *A Late Awakening*, Chapter 5; *Numbers Decide*, 191-196; *Savior of Science*, 1-8. For lengthier treatments, most developed in his later writings reflective of his life's work, see *A Mind's Matter*, Chapter 14 and Stanley L. Jaki, *Means to a Message: A Treatise on Truth* (Grand Rapids, MI: William B. Eerdmans Publishing Company, 1999), Chapter 3.

[24] Jaki, *Science and Religion*, 5.

[25] Jaki, *Science and Religion*, 28.

will. This is the theological explanation for the intellectual power to even do science and the will to do it right, which is why Jaki highlighted the "power" of reasoned discourse. More simply, humans innately know what numbers are, but numbers are not the limit of all human reasoning.

So then, the other sciences besides physics are all exact insofar as they are the application of quantities to objects by measurement. To the extent that the various branches of science use numbers connected with measurements, they are *exact*, but that distinction does not, as shown before, disallow for non-exact sciences. The attentive reader may have guessed by now where this is headed and why Jaki's concise definition based on quantities and exactness may be rejected by those who adhere to Scientism. The logical conclusion points to the difference in religion and science:

> One should not forget even for a moment the most fundamental rule (indeed a fact) which should govern all reasoned (indeed sane) discourse about the relation of science and religion: Quantities form one conceptual domain, and all other concepts another. Painful as this may be to incurable reductionists, the two domains are and remain irreducible to one another.[26]

To grasp how quantities form a conceptual domain independent of all other concepts is to grasp that Jaki clarified a fundamental distinction between science and religion that, one can speculate, should have been heeded more carefully all along in the history of science. The distinction has still not been fully developed, but hopefully these ideas will lead to more clarified discussions about the future of science.

---

[26] Jaki, *Science and Religion*, 28.

Why Does This Definition Matter?

Why Does This Definition Matter?

Scientists today have difficulty defining their own field because of the failure to distinguish exact science from reasoned discourse, and if scientists cannot even define science, then neither can anyone else. There is a very real difficulty and confusion, though students of science will often not admit it. Maybe they are not aware of the difficulty because they were never taught it, or maybe they want science to have more power than it does. The *Academic Press Dictionary of Science & Technology*, for instance, defines science as "the systematic observation of natural events and conditions in order to discover facts about them and to formulate laws and principles based on these facts."[27] But what does that mean, and where are the boundaries?

Jaki took issue with turning science into philosophy, something such a vague definition certainly allows since "events," "conditions," and "principles" can be interpreted broadly.[28] Jaki noted that even such giants of modern science, like Max Planck, Albert Einstein, Werner Heisenberg, Niels Bohr, Erwin Schrödinger, and Arthur Compton produced books that gave credence to the erroneous perception that exact science is a "potent source of philosophical insight."[29] If Jaki's insistence that exact science be limited to measurement of objects in motion was followed, it would be clear to anyone that science cannot answer all the problems that humans face. It would be clear that the desire to graft philosophy onto something non-philosophical (or non-science onto what is scientific) is a perversion of both science and philosophy that ought to be avoided for the sake of true progress.

---

[27] Christopher G. Morris, *Academic Press Dictionary of Science & Technology* (San Diego, CA: Academic Press, 1992), "Science."

[28] Jaki, *A Mind's Matter*, 3, 96, 247, 255.

[29] Jaki, *Means to a Message*, 75.

And so, in an even deeper sense, what Jaki really did with his insistence that science, the *quantitative* study of the quantitative aspects of *objects* in motion, be grounded in exactness was to protect science *as* science so that other questions of humanity, existence, philosophy, and religion could be considered more fully with the power of reasoned discourse. That these disciplines have been marginalized by an over-reliance on "science" to shore up positions that are actually the work of the intellect separate from numbers and quantities is proof that such perversion has already occurred.

Solid reasoned discourse, including philosophy and theology, does not need quantification from exact science to support it, since those discourses ought to be able to stand on their own merit. That is a most significant point. It is strong praise of the human ability to reason, an ability that is not praised enough or given enough confidence by a modern culture that too often invokes science as if it were a marketing tool. Understanding this distinction makes it possible to wade through the immense studies and claims wrapped up as science, or as Jaki loved to say, used as one of the three S's, Sport, Sex, and Science (and sometimes he added a fourth, Smile), that sells ideas.[30]

Consider the repercussions of having confused these distinctions. In 2009, the Science Council in the U.K. announced a year-long effort to give a new "official" definition of science, the first "ever published" according to the *The Guardian* periodical.[31] It took a year for the members of the council to declare science to be "the pursuit of knowledge and understanding of the natural and social world following a systematic methodology based on evidence." This

---

[30] Jaki was fond of this saying and repeated it often.

[31] Ian Sample, "What is this thing we call science? Here's one definition . . ." at *The Guardian News and Media* (3 March 2009) at http://www.theguardian.com/science/blog/2009/mar/03/science -definition-council-francis-bacon.

definition may sound sophisticated at first, but, just as is the *Academic Press* definition does, it leaves room for philosophy-grafting. As such it gives science an undue, false power to describe the universe and mankind's purpose. It is remarkable that the atheist philosopher A.C. Grayling applauded this definition for its generality.[32] The danger with these ambiguities in an increasingly secular culture is that under such broad and indistinct definitions, it is possible for non-religious people to conclude that science can give the answers to non-scientific questions that have nothing to do with quantities and matter, ergo Scientism. As Jaki said, it leads to the conclusion that science is the "savior of mankind."[33]

One can only wince at what Jaki would have said about the essay contest organized and funded by philosopher and neuroscientist Sam Harris, who wrote a book in 2010 titled *The Moral Landscape: How Science Can Determine Human Values*, which claims that "questions of morality and values must have right and wrong answers that fall within the purview of science."[34] In 2014, Harris is offering $2,000 to the winner of an essay contest to refute the premise of his book and $20,000 to anyone who convinces him to abandon the premise. Jaki undoubtedly would have turned out a clever phrase about the contradictions in moral landscapes, human values, and monetary value being grafted onto a mind who also wrote the book that claims free will is an illusion.[35] Indeed people have looked upon Harris' claim about science determining human values with consternation and rejection, but so lost has modern thought become in the absence of clarified distinction between *quantities* and *qualities* that

---

[32] Sample, "What is this thing we call science?"

[33] Jaki, *A Mind's Matter*, 52, 57.

[34] Sam Harris, *The Moral Landscape: How Science Can Determine Human Values* (New York, NY: Free Press, 2010).

[35] Sam Harris, *Free Will* (New York, NY: Free Press, 2012).

opponents are at a loss to understand why Harris could be wrong.

To even say something simple such as "a rock *is* there" is beyond the purview of exact science. How shocking is that? But to ponder it reveals something quite profound. Quantities do not alone identify what something *is*. Science cannot define what a substance *is*, or whether the substance is an object or a being. Science does not measure demonstrative pronouns or identify verbal states of being, a radically simple concept to grasp but not a concept commonly acknowledged. Science only addresses the *how*.[36]

Exact science is therefore extremely limited in its applicability. Jaki thought this was a standard of most importance, not just for scientists, but even more so for philosophers and theologians. Consider this strong admonishment: "Whenever a philosopher/theologian yields the fraction of a hairsbreadth on the intrinsic limitation of exact science, he runs the grave risk of the heedless boy who put his hand through the fence of the lion's cage. The risk is that of a potential tragedy."[37] The advice there is of utmost importance for theologians. Theologians do not need to rely on science to shore up the material of their discipline. They do not need information about quantities of objects in motion to defend what reason can discover or what God has revealed. Scientific discovery can contribute to the appreciation or understanding of philosophy and theology, but those disciplines ought to first stand on their own merits. So much for science, as Jaki would say.

In an essay given in 2003, "What God Has Separated . . ." Jaki taught that this separation of science (quantities) and religion is not just a separation of his making, but one God Himself created in the mind of man. Mankind knows this separation because the human mind innately knows the

---

[36] Jaki, *A Late Awakening*, 66.
[37] Jaki, *Science and Religion*, 6.

difference between quantities and everything else. Christ demonstrated this too when asked about what to do with a coin. The Lord said, "Give back to Caesar what is Caesar's, and to God what is God's."[38] Purpose and free will are qualities that cannot be measured, and therefore, their consideration does not belong to science or to things limited to this physical world. They are matters for revealed religion or philosophy. The reminder in Mark's Gospel is a good one for those who think science can explain everything. "How is a man the better for it, if he gains the whole world at the expense of losing his own soul?"[39] Truly, a person could learn all the science there ever could be learned about the motion of objects in the physical world, and it would not save his soul.

"Science does not tell us what we should do, it does not even tell us what *is*, simply because there are no units of measurement for *is*. Revealed religion depends and rests on that verb *is*."[40] God is existence itself, "I Am who Is," and religion is about purpose of the highest conceivable kind, life after death and personal immortality.[41] The etymology of the word "religion" means to *re-ligare*; *ligare* means to tie a knot, so *re-ligare* means to reunite with God.[42] Jaki made this point often, the only real religion is the one "steeped in Christian Revelation," a personal Creator who brings forth the universe out of nothing and who inspired man to pray to Him by becoming Incarnate and by purchasing through his life, death, and resurrection the eternal rewards of Redemption. Jaki summed it up beautifully: "I only wish that Catholics would

---

[38] *The Knox Translation Bible* (Westminster: Baronius Press Ltd., 2013), Mark 12:17.

[39] Jaki, *A Late Awakening*, 68; Knox, Mark 8:36.

[40] Jaki, *A Late Awakening*, 63, 66.

[41] Jaki, *A Late Awakening*, 67.

[42] Jaki, *A Mind's Matter*, 13.

really cherish the word *is* even though science cannot say anything about it."[43]

Jaki mentions in his autobiography that he had the goal to more fully develop this idea in his 1999 publication, *Means to a Message: A Treatise on Truth*. He explains that "the scientist singles out what is quantitative in reality and therefore he deals only with matter and only insofar as matter embodies quantitative features."[44] The scientist, *qua* scientist, does not probe into the deeper questions about where the matter came from, why it is here, or who gave it its quantitative properties; he instead accepts that those objects really exist *a priori*.[45] Those questions about existence and being belong to philosophy and theology. This insight is valuable because it puts science at a dead stop with the material world and measurement.

In the final chapter of his intellectual autobiography, Jaki further wonders why it was that even Aristotle failed to more methodically treat one of his "most portentous remarks."[46] Jaki first remarked on this idea thirty years prior in *The Relevance of Physics*, and he spent much time developing it because it applies to his argument about "The Savior of Science." One of Jaki's greatest contributions to the understanding of science is based on this simple concept, the role which quantities play in science, a concept overlooked in ancient times as well as in modern, a concept Jaki himself realized he had not fully appreciated in his earlier works.

If there is a conviction that has grown in me during these last four or five decades, it relates to something with very sharp edges and well-defined faces. I mean the decisive role which quantities play in science and their inability to

---

[43] Jaki, *A Late Awakening*, 70.
[44] Jaki, *Means to Message*, 7.
[45] Jaki, *Means to Message*, 7.
[46] Jaki, *A Mind's Matter*, 242.

play that role elsewhere. On re-reading various books of mine on science, I find more than one proof of my having been aware of this difference early on.[47]

The significance of this simple rule is that it at once demonstrates why science cannot be un-scientific. Religion is about morality; science is not. It also subordinates science to religion because cultures need the former more than the latter. In his autobiography, Jaki explains that "real culture" must attend to questions that "most agitate a human being."[48] There must be attention to religion's questions, and since religion cannot exist without the form of a cult, worship directed towards God, *cult*ures–by necessity, by definition– must do so. As any admirer of Jaki's knows, this is where his brilliance and his passion united with force. He could speak or write so powerfully about science and culture and point his audience straight to Christ.

> Real cult means real religion, that is, a religion with a God in its center to whom man can be truly "re-ligated" (the etymology of the religion) so that he may truly worship. No true worship is deserved by a God who is the product of a cosmic process, let alone the distillation of a process theology. The only God who deserves a proper cult, which is worship, is much more than the Creator who brings forth the universe out of nothing.[49]

---

[47] Jaki, *A Mind's Matter*, 242.
[48] Jaki, *A Mind's Matter*, 13.
[49] Jaki, *A Mind's Matter*, 13.

Chapter 2 – "Was Born"

Chapter 2 – "Was Born"

Jaki did not argue that science was invented by Christians or that it originated with Christians. He argued that science, as defined in the previous chapter, was born into a "viable discipline" from a "cultural womb" that properly nurtured it, and that it was nearly born in other cultures too, but died. By "was born" Jaki means a transition–a *breakthrough*–from independent insights, skills, and practical inventions to a self-sustaining discipline of theoretical generalizations and investigations about the physical world. Science, once born, matured into a universal enterprise of exact physical laws and systems of laws.[1]

As outlined in the Introduction, this chapter is a four-part review based on, but not limited to, the information provided in *Science and Creation*. First, the scientific successes of each culture are described and then the reasons for the "stillbirths" of science within each one. Second, the Old Testament worldview and how that biblical view purified a truly scientific worldview is discussed with ample quotation. Third, a brief presentation of the scientific attitudes of the early Church Fathers is given to show the continuity from the biblical cultures into the Middle Ages. Fourth, a lengthy section that addresses the contributions of ten medieval Christian scholars is given in chronological order to show how the progression toward a breakthrough in scientific thought was based on divine revelation. This is by far the longest chapter of the book, and it covers a lot of history but highlights the aspects relevant to Jaki's research.

---

[1] Jaki, *Science and Creation*, 14.

Stillbirths in Ancient Cultures

Stillbirths in Ancient Cultures

"In the cycle of existence I am like a frog in a waterless well."[2]

History is a curious discipline because although the researcher is searching for facts, the facts are collected by interpreting what the original writer meant to convey. Jaki preferred original sources to secondary ones for that reason.

> There is no substitute to the perusal of primary texts, which, incidentally, hardly ever fail to reveal something that has not yet been noticed by others. Engrossment with the secondary literature can readily trap one's vision along tracks that have little in common with the thrust of the primary sources.[3]

Since this is a sifting through of Jaki's writings on this question about the birth of science, it was tempting to merely repeat what Jaki wrote, but that would have violated the above warning of potential entrapment. Therefore, throughout this chapter, original sources were found, read, quoted, and cited, except for the small amount of instances where footnoted.

Haffner noticed that Jaki even described in *The Relevance of Physics* that he saw the development of science as an "ongoing

---

[2] *Maitri Upanishad*, First Prapathaka, last line. Another translation reads: "In such a world as this what have I to do with the enjoyment of desires? Yea, Even if one were fed therewith to the full, he must still return to earth again and again. Wilt thou therefore deign to deliver me? I am here in this world as a frog in a well without water. Oh adorable one, thou art our refuge, thou art our refuge." *Maitri Upanishad, Sanskrit Text with English Translation*, edited and translated by E. B. Cowell (London: Asiatic Society of Bengal, 1870), 244; Jaki, *Science and Creation*, 7.

[3] Jaki, *A Mind's Matter*, 5-6.

process" that is something "genuinely human" and a "mixture of achievements and failures characterized by incessant changes."[4] Jaki understood that science (even before it was called science) evolved throughout human history. When Jaki referred to the "birth" of modern science (or exact science), he was referring to the application of mathematics (quantification) to physics (objects in motion), the change from classical physics to Newtonian physics, or what some call the Scientific Revolution.[5]

When Jaki coined the term "stillbirth of science" in reference to the Greek, Egyptian, Babylonian, Hebrew, Muslim, Chinese, and Hindu cultures, he was, therefore, not implying that there was never any progress or breakthroughs in science. He was referring to a specific, but vitally significant, breakthrough. He used that phrase "stillbirth" to demonstrate that no culture before the Christian culture made the leap to measure changes in material objects with quantities in a systematic and self-sustaining "viable" way. Here is how Jaki put it:

> It is not a pleasant task to call attention to the obvious. To make others appear to be shortsighted, let alone blind, may easily evoke resentment. But it had to be obvious and clearer than daylight that in none of those cultures, although they lacked no talent and ingenuity, did science become a self-sustaining enterprise in which every discovery generates another. In all those cultures the scientific enterprise came to a standstill. It is this phenomenon which I called the stillbirths of science.[6]

---

[4] Haffner (1991), 21; Jaki, *The Relevance of Physics* (Chicago, IL: University of Chicago Press, 1966), 502.

[5] Haffner (1991), 21.

[6] Jaki, *A Mind's Matter*, 52.

In the 1986 volume of 377 pages, *Science and Creation: From Eternal Cycles to an Oscillating Universe*, Jaki presented his historical research for these stillbirths.[7] In the briefer book *The Savior of Science*, he summarized these stillbirths in the first chapter.[8]

The history of science has been presented to modern students (if the history is taught at all) as a linear climb out of darkness, a development of mankind that heightened him to where he is today. This is how the history, for example, was set forth in the 1904 five-volume account, *A History of Science*, by Henry Smith Williams and Edward Huntington. These authors held that for all the history of mankind, science was developing neatly in a linear fashion, one step necessarily leading to the next, culminating to the present day.

> We shall best understand our story of the growth of science if we think of each new principle as a stepping-stone which must fit into its own particular niche; and if we reflect that the entire structure of modern civilization would be different from what it is, and less perfect than it is, had not that particular stepping-stone been found and shaped and placed in position. Taken as a whole, our stepping-stones lead us up and up towards the alluring heights of an acropolis of knowledge, on which stands the Temple of Modern Science. The story of the building of this wonderful structure is in itself fascinating and beautiful.9

It is significant to note that Jaki rejects this faulty, but popularly received, stepping-stone, linear historical model for

---

[7] Jaki, *Science and Creation*.

[8] Jaki, *Savior of Science*.

[9] Henry Smith Williams and Edward Huntington Williams, *A History of Science* (New York: Harper & Brothers, 1904), Volume 1, Book I, Introduction.

the development of science. Not only would such a stepping-stone model be impossible before the invention of global communication, 1904 obviously did not turn out to be the pinnacle for science they claimed it was. Instead Jaki described the history of science more naturally, as an evolutionary tree that had many dead branches before it flourished into a vital and self-sustaining, living discipline.

Jaki also described the history of science as a theological history. Science is a work of mankind, and as such, it must take into account different cultures and cultural mindsets. Since every culture in the history of man sought God, the evolution of science must take into account the religions of those cultures too and ask some "searching questions."[10] This is why Jaki takes the analogy another step further in arguing that empirical science—exact science based on quantities, as discussed before—was born of Christianity, which means it was born of a woman, a Virgin Birth. Jaki referred to the dead branches, as it were, on the evolutionary tree as stillbirths, as living entities that developed for a while in a womb (a certain culture) but died before becoming viable as a universal discipline recognized for its own methods.[11]

Science and Creation was not just about the history of civilizations, it was about the history of science "in its relation to civilizations, ancient and modern."[12] The book centered on the role of religion in civilizations, exhaustively citing prose and verse to confirm the theological convictions that hindered the viable birth of science. *Science and Creation* treats ancient cultures/periods in this order: India (Hinduism, Buddhism); China (Taoism, Moism, Confucianism); the Aztec, Inca, and Maya civilizations in pre-Columbian American; Egypt; Mesopotamia and Babylon; Greece; Hebrew (Judaism); the Roman Empire (early Christianity);

---

[10] Jaki, *Savior of Science*, 22.
[11] Jaki, *Savior of Science*, 20-22.
[12] Jaki, *A Mind's Matter*, 48.

Arabia (Muslim and Islam); European Middle Ages; the Renaissance; Galileo and Kant; Romanticism; and Modern Science.

As mentioned previously, the *Science and Creation* volume contains 377 pages, much of which contains references to original and secondary documents. Jaki built the case that was forming in his mind then and crystallized in his later works: There was no shortage of cultures ripe with skill, talent, and yearning in which a scientific birth could have taken place, yet it did not until the Christian Middle Ages, not so much by any monumental breakthrough in ability, but by acknowledging a rational world, a rational mind of man, and an application of mathematics to objects in the real world, a world with an absolute beginning and end of time. While researching this book, Jaki began to realize that the only "viable birth" of science took place in a matrix "that was not merely monotheistic, but also Christologically monotheistic."[13]

> Meanwhile I took immense delight in enlarging my mind by delving into ancient Hindu, Chinese, pre-Columbian American, Egyptian, Babylonian, and Greek history, into a re-reading of biblical history and of patristic literature. They contained many indications that I was on the right track in unfolding a new vision of the history of science, not attempted beforehand. Some of the indications were a sheer delight to stumble on.[14]

He concluded the book with a speculation that if the psychology of the modern world returns to the psychologies of past cultures, symbolized by the treadmill of eternal cycles, science will not survive.

---

[13] Jaki, *A Mind's Matter*, 54-55.
[14] Jaki, *A Mind's Matter*, 55.

In *The Savior of Science,* Jaki only devoted the first chapter, which is forty-eight pages long, to the "stillbirths" of science, and it is a summary of the work found in *Science and Creation.* That chapter, however, only considered six ancient religions and civilizations, following a commentary on the progress of science and a commentary on what Jaki means by "stillbirths" of science. He addressed the religions and civilizations in a different order than in *Science and Creation*: Egypt, India, China, Babylon, Greece, and Arabia. The reason for this order appears to be in terms of the religion, first the pantheistic religions which did not communicate with each other and therefore did not form a "critical mass of knowledge resulting in a chain reaction, or rise to the intellectual temperature where self-ignition takes place" (Egypt, India, and China) and then the polytheistic and monotheistic ones which developed in "direct succession" of one another (Babylon, Assyria, and Persia to the Greece, and from them to the Arabia).[15] His purpose in that book, and elsewhere, was to further the argument, an "entirely original" argument, that the birth of Christ was "The Birth that Saved Science."[16]

For the remainder of this section, the cultures taken up in *The Savior of Science* will be reviewed, but not exactly as presented in that book. The review, having the benefit of Jaki's hindsight offered in his intellectual autobiography and the benefit of wide accessibility to resources through the internet, will include some of the references Jaki used in both books, along with some other resources to support his thesis, with a preference for the examples that would be most beneficial to someone arguing this claim in current times.

The difficulty in making this argument that science was born of Christianity is that anyone who is unfamiliar with the revolutionary historiography put forth in *Science and Creation*

---

[15] Jaki, *Savior of Science*, 37.
[16] Jaki, *A Mind's Matter*, 66.

will obscure the conclusion by pointing out all the contributions to science that other cultures made, contributions that Jaki does not ignore. No legitimate criticism can be made unless Jaki's detailed work in *Science and Creation* has actually been read, considered, and if desired, verified. What follows are the historical details useful in understanding his approach. The scientific successes of each culture will be described and then the reasons for the stillbirths of science within each one, as Jaki saw it, will be presented. What will emerge is the meaning of the opening quote, that all stillbirths were owed to a worldview that the whole universe, including man, is an eternal cycle of no escape.

Egypt

The first stillbirth Jaki discussed in the *Savior of Science* is the stillbirth of science in Egypt, "an Egypt to be buried in the sand."[17] In ancient Egypt (from about 3000 B.C.), impressive discoveries and achievements were recorded in history. The Egyptians constructed grand pyramids of such majesty and awe that no one today knows how they did it. They invented hieroglyphics, a highly developed form of phonetic writing which may have been the greatest intellectual feat of its kind. They had medical arts. They were successful in using the Nile as an abundant resource. They adopted better weaponry and the use of chariots from other countries. The Egyptian king, Wehimbre Neco, who ruled from 610–595 B.C., sent a fleet to sail West, and the sailors traveled from the Arabian gulf into the southern ocean for three years until they returned to Egypt.[18]

---

[17] Jaki, *Savior of Science*, 22-25.
[18] Jona Lendering, "The First Circumnavigation of Africa," Moellerhaus at
http://www.moellerhaus.com/Persian/Hist01.html.

Egyptian social life revolved around practical skill. For the proper distribution of grain and other commodities, ancient Egypt relied on a system of arithmetic in which they took stock of and divided out resources with impressive book-keeping skills.[19] They invented a decimal system with special glyphs for powers of ten up to one million. Their calendar endured uninterrupted use during all of Egyptian history, and the Hellenistic astronomers adopted it for their calculations.[20] Ptolemy based his tables on this calendar in the Almagest on Egyptian years, as did Copernicus to some extent.[21]

Ancient Egyptian craftsmen showed great ingenuity in using their tools. They had a simple but effective method of producing sheets of paper from the leaves of the papyrus plant, much more efficient than the use of animal skin as writing substrates.[22] They were the first to produce plywood as many as six layers deep and made of mixtures of woods. Carpentry among Egyptians used methods of joining wood in intricate patterns for the hulls of boats as well as inlaying, veneering, and overlaying techniques.[23] The burial chambers of Pharaohs of the XVIIIth Dynasty from the sixteenth century B.C. have received much publicity for their highly developed architectural planning containing secret chambers that even space-age technology and sensitive cosmic-ray methods could not detect.[24]

The pyramids, however, constitute the real mystery in Egyptian marvel and ability. Their proportions were enormous. The Egyptian stonecutter placed the huge blocks

---

[19] Jaki, *Creation and Science*, 68.
[20] O. Neugebauer, "The Origin of the Egyptian Calendar," *Journal of Near Eastern Studies* (1942), 396.
[21] Neugebauer, 396.
[22] Jaki, *Science and Creation*, 79.
[23] Jaki, *Science and Creation*, 79.
[24] Jaki, *Science and Creation*, 79.

of stone together with only 1/50 of an inch separation at the base of the pyramid and covered them with marble plates of such smoothness that the pyramids looked like mirrors.[25] They managed to quarry, shape, and polish great stones despite the fact that they had no metal tools. Transportation of the great stones was done with wooden sleds. The overall master plan of the pyramids formed a superbly constructed facility to ensure the king's journey to the Sun God.[26]

Even with these achievements, the underlying theology and *cultural mindset* regarding the universe thwarted scientific advancement. "In their deepest meaning the pyramids were symbols of a conception about the world that nipped in the bud all scientific endeavors."[27] The Egyptians were caught up in an animistic, cyclic outlook that made them insensitive to science as well as history. In their hymns they pictured most parts of the world as animal gods, the whole world itself being one huge animal often depicted as a serpent bent into a circle. In a hymn from ancient texts, the animistic, organismic, rhythmic, and cyclic worldview is explicitly described:

> He [the Indwelling Soul] it was who made the universe in that he copulated with his fist and took the pleasure of emission. I bent right around myself, I was encircled in my coils, one who made a place for himself in the midst of his coils. His utterance was what came forth from his own mouth.[28]

---

[25] Jaki, *Science and Creation*, 79.

[26] Jaki, *Science and Creation*, 79.

[27] Jaki, *Science and Creation*, 79.

[28] R. T. Rundle Clark, *Myths and Symbol in Ancient Egypt* (London: Thames & Hudson, 1959), 51; quoted in Jaki, *Science and Creation*, 73.

The Egyptians believed that the circularity in the sky and in nature was proof that the cosmos was changeless and cyclical too, and that single events or processes had little or no significance, which meant that they "simply could not serve as the carriers of special intellectual content."[29]

The Egyptians had the talent and the skill to notice that everything in the material world is in motion and is, thus, observable and quantifiable. They had the talent to realize that the scientific method could be applied repeatedly to answer questions about the universe, to determine scientific laws. They had the ability to innovate and the ability to communicate it. They demonstrated the ability to learn from other cultures. Science could have been born in ancient Egypt, but it was not. All of that progress came to a standstill, a stillbirth.

Jaki also pointed out that to argue that "the Egyptians of old failed to develop more science because they did not feel the need for more is an all too transparent form of begging a most serious question," a conceited psychology.[30] If they had been but an animal species, they would have never even tried to innovate. They would have continued on their way with things as they were, just as all other animals do. There was plenty of evidence that they did long for something better. During the reign of Akhenaton, the Pharaoh known for abandoning traditional Egyptian polytheism and introducing worship of Aten, a monotheistic religion's view, Egyptians responded in great number to dispose of long-established rigid art forms and seek "warmly humane representations of life and nature."[31] Egyptians seemed to want something better.

---

[29] Jaki, *Science and Creation*, 72.
[30] Jaki, *Savior of Science*, 23.
[31] Jaki, *Savior of Science*, 23.

Yet after Akhenaton's death the traditional religion was restored and Akhenaton became archived as an enemy.[32] The longing is evident in the poetry the Egyptians sang, the inspiration they took from the animal kingdom in their carvings of animal and human combined bodies, effigies which now are, as Jaki put it, "buried in the sand as if to symbolize that there was no future in store for the Egypt of old."[33] In a culture of pantheism, where the people saw themselves as part of an animate universe, modern science could have been born, but was not. Eternity consisted in assimilating to the cyclic motion of nature; souls that reached the stars were considered transfigured spirits absorbed into the great rhythm of the universe.[34]

China

There is so much written about China's rich and illustrious past that no case could ever be made–from the Shang Dynasty (1523–1028 B.C.) to the Ch'ing Dynasty (A.D. 1644–1912) that there was no progress in civilization, art, or literature. Likewise, volumes have been written on the question of the history of science and Chinese civilization. In *Science and Creation* and *The Savior of Science*, Jaki referred to the extensive research of British biochemist Joseph Needham.[35] In seven volumes comprised of twenty-seven books, Needham and his team of international collaborators reviewed the history of science and technology in China. The massive work was eventually published by the Cambridge University Press under the title *Science and Civilisation in China*;

---

[32] Jaki, *Savior of Science*, 24.

[33] Jaki, *Savior of Science*, 25.

[34] Jaki, *Science and Creation*, 72.

[35] Jaki, *Science and Creation*, 46; Jaki, *Savior of Science*, 35.

and the project, which began in 1954, continues to the present day.[36]

A brief overview of the content of these volumes will demonstrate the extent of cultural development in China and the futility of ignoring such a rich history. Needham's first volume (1954) is an introduction to the rest of the work. Volume Two (1956) covers the history of scientific thought in China, including the organic naturalism of the great Taoist school, the scientific philosophy of the Mohists and Logicians, and the quantitative materialism of the Legalists.[37] Volume Three and the three-part Volume Four (1959–1971) addresses mathematics and the sciences of the heavens and earth, physics, mechanical engineering, civil engineering, and nautics. Volume Five (1985–1999) has thirteen parts: the first on paper and printing; the second through the fifth on spagyrical discovery and inventions including gold and immortality, cinnabar elixirs, synthetic insulin, apparatus, and physiological alchemy; the sixth and seventh on military technology from missiles and sieges to the "gunpowder epic;" the ninth on the textile industry while the eighth and tenth are still works in progress; the eleventh on ferrous metallurgy; the twelfth on ceramic technology; and the thirteenth on mining. Volume Six (1986–2000) deals with botany, agriculture, agroindustry, and forestry in the first three parts and fermentations, food science, and medicine in the fifth and sixth parts, while the fourth part of Volume Six is still in progress. Finally, Volume Seven (1998–2004) covers language and logic, and then gives the general conclusions and

---

[36] See the Needham Research Institute website, http://www.nri.org.uk/.

[37] Joseph Needham, *Science and Civilisation in China: Volume 2, History of Scientific Thought* (Cambridge, UK: Cambridge University Press, 1956).

reflections.[38] The purpose of listing these volumes published over a span of six decades is to demonstrate that intensive work has been devoted to the history of science in China, and Jaki was aware of this. He acknowledged it in the development of the "stillbirths" argument.

In *Science and Creation* Jaki discussed how around 350 B.C., the astronomer Shih Shen drew up a catalogue of around 800 stars and how the manuscripts were stored in the Imperial Library.[39] The ability to catalogue and store documents displayed great sophistication. Technological improvements were made in water works and the extension of the Great Wall, a massive achievement. During the three and a half centuries known as the age of the Warring States (480–220 B.C.), cultural growth continued. The Chinese invented the waterwheel, the wheelbarrow, and other devices that demonstrated continued technological development.[40] Around the middle of the fourth century, Hu Hsi made observations that led him to discover the precession of equinoxes, although the Greek scholar Hipparchus is credited with discovering it centuries earlier.[41]

The peak periods of Chinese culture spanned the Han, Sung, Thang, Yuan, and Ming periods (collectively 202 B.C.–A.D. 1644) and represented a length of time when scientific endeavor could have "received a decisive spark."[42] There were technological feats in which the Chinese were the "sole inventors" for a number of centuries. They invented the effective use of horses, the foot-stirrup and breast-strap harness. They discovered magnetic ore. They invented the revolutionary skill of paper-making, which led to the

---

[38] A full list may be obtained at the Needham Research Institute website, http://www.nri.org.uk/science.html.

[39] Jaki, *Science and Creation*, 32.

[40] Jaki, *Science and Creation*, 32.

[41] Jaki, *Science and Creation*, 32.

[42] Jaki, *Science and Creation*, 32.

production of printed books. They invented the process of making gunpowder, the production of porcelain, and the development of water-driven mechanical clocks. They used magnets for travel and moveable clay types for printing.[43]

The Chinese also, Jaki noted, developed algebra at a level compatible with the best in Europe around A.D. 1250. According to Francis Bacon, printing, gunpowder, and magnets were the factors that ushered in the age of science more than anything, but Jaki challenged Bacon's assertion by noting that even with these developments the Chinese "remained hopelessly removed from the stage of sustained, systematic scientific research."[44]

The Chinese had rockets for centuries but did not investigate trajectories or free fall. Their ability to print books did not lead to a "major intellectual ferment." Magnets were installed on their ships and they were the best navy in the world for the fourteenth and fifteenth centuries, but they never circumnavigated the globe.[45]

Historians have also noted that the "Industrial Revolution" did not originate in China, and that is of great significance for Jaki's argument that science was "stillborn" in Chinese culture. Jaki cited a 1922 article in *The International Journal of Ethics* entitled "Why China Has No Science: An Interpretation of the History and Consequences of Chinese Philosophy."[46] The author, Yu-Lan Fung, who contributed to Needham's volumes, noted that the history of Europe and the history of China before the Renaissance are "on the same level," by which he meant that they both progressed at about the same pace, albeit in different ways. After that time the

---

[43] Jaki, *Science and Creation*, 32.

[44] Jaki, *Science and Creation*, 32.

[45] Jaki, *Science and Creation*, 32.

[46] Yu-Lan Fung, "Why China Has No Science: An Interpretation of the History and Consequences of Chinese Philosophy," *The International Journal of Ethics*, 32 (1922), 237-263.

pace differed: "China is still old while the Western countries are already new."[47] Fung asked, "What keeps China back?" He answered that it is because "she has no science . . . because according to her own standard of value she does not need any . . . China has not discovered the scientific method, because Chinese thought started from mind, and from one's own mind."[48] If truth and knowledge are in the mind, separated from the external world, there is no need for scientific investigation beyond practical skill.

Fung contrasted the three major powers which competed to conquer the entire empire of China from 570 B.C. to about 275 B.C–Taoism, Moism, and Confucianism.[49] Taoism taught a "return to nature" with nature being the natural state of all things, including the natural tendency of man toward vic. According to Taoism, "every kind of human virtue and social regulation is to them against nature."[50] Knowledge was considered to be of no use because the Tao is inside man, as the god of the pantheistic philosophy.[51] Taoism did not require any questioning of a beginning and an end, about final purposes and goals, or about the controlling of the forces and patterns in the workings of the Yin and Yang.[52] The cosmological passage from the *Chuang Tzu* demonstrated this mindset: "Men who study the Tao do not follow on when these operations [properties belonging to things] end, nor try to search out how they began: - with this all discussion of

---

[47] Fung, 237-238.
[48] Fung, 238.
[49] Fung, 240.
[50] Fung, 241-242.
[51] Fung, 243.
[52] Jaki, *Science and Creation*, 30.

them stops."[53] The key to success in Taoism was to merge into the rhythm of cosmic cycles.[54]

The fundamental idea of Moism was "utility," and virtue was seen as useful.[55] Universal love was taught as a doctrine for the benefit of the country and people, and progress was the ideal of mutual help; anything that was incompatible with the increase of wealth and population was to be fought against.[56] Confucius stood between the two, emphasizing discrimination in different situations.[57] He taught that human nature is essentially good although men are not born perfect.[58] To become perfect, the innate reason must be developed and lower desires "wholly taken away."[59] His concerns were ethical, not metaphysical. Therefore, Confucius taught that the individual should seek what is in himself and leave external things to their natural destiny.[60]

In these competing theories of existence, the power that governs the universe is the omnipotent Tao for Taoism, the personified self-god in Moism, and Heavenly Reason according to Confucianism.[61] Moism did have a notion of Heaven as personal and caring for humans, a monotheism of sorts, but its ethics were severed from this idea. As these powers competed over time, to put it far too concisely to do the history enough justice, they actually merged and philosophical investigation of "things" gave rise to two forms of Neo-Confucianism, one school that sought "things"

---

[53] Texts of Taoism, translated by J. Legge (New York: Julian Press, 1959), Book XXV, par. 11, 568-69; quoted in Jaki, *Science and Creation*, 30.

[54] Jaki, *Science and Creation*, 27.

[55] Fung, 244.

[56] Fung, 245-248.

[57] Fung, 249.

[58] Fung, 250-252.

[59] Fung, 252.

[60] Fung, 253.

[61] Fung, 254.

externally and another that sought "things" as phenomena in the mind.[62] In Medieval Europe the same ideas about "things" more or less existed too, but from there on, China and Europe diverged:

> In other words, Medieval Europe under Christianity tried to know God and prayed for His help; Greece tried, and Modern Europe is trying to know nature and to conquer, to control it; but China tried to know what is within ourselves, and to find there perpetual peace.[63]

So China did not have use for the scientific method because the religions sought what is in the mind separate from the external world. Fung concluded his paper with a call for mankind to become wiser and to find peace and happiness by turning attention to Chinese wisdom so that the "mind energy of the Chinese people of four thousand years will yet not have been spent in vain."[64] Even if modern science was not born in China, there were other aspects of the culture that were worthy of admiration.

In concluding this consideration of China's history, it needs to be noted that other scholars concurred with Fung. In 1995, Justin Yifu Lin of Peking University published an essay titled "The Needham Puzzle: Why the Industrial Revolution Did Not Originate in China." Lin noted from evidence documented in Needham's work that "except for the past two or three centuries, China had a considerable lead over the Western world in most of the major areas of science and technology."[65] From an economic and social perspective,

---

[62] Fung, 258.

[63] Fung, 259.

[64] Fung, 263.

[65] Justin Yifu Lin, "The Needham Puzzle: Why the Industrial Revolution Did Not Originate in China," *Economic Development and Cultural Change* (University of Chicago Press, 1995), 269-292.

he considers why, despite early advances in science, technology, and institutions, China did not take the next step in the seventeenth century as Western Europe did.[66]

Ultimately that answer depends on how the Chinese viewed the external world and whether it was created by God or was God itself. In believing that the world *was* God and was eternal, there was no need to question a beginning and an end or how everything came to be. Even Needham acknowledged that it is a *theological* orientation of Chinese thought that can be singled out as the decisive factor that blocked the attitude conducive to developing a systematic, scientific investigation.[67] "There, according to Needham's admission, all the early cultivators of science drew courage for their pioneering efforts from a belief in a personal and rational Creator."[68]

For the purposes of Jaki's argument, the similarity of the Egyptian and Chinese cultures thus considered bears emphasizing. Both were pantheistic, with some degree of monotheism but still a monotheism that held that the world was God, which is basically pantheism. Neither had a loving Creator who "ordered all things in measure, and number, and weight," who made man in His image with intellect and free will, or who became Incarnate to redeem mankind.[69] "In a universe without the voice of God there remains no persistent and compelling reason for man to search within nature for distinct voices of law and truth."[70]

India

---

[66] Lin, 271.

[67] *Science and Civilisation in China: Volume 2, History of Scientific Thought*, 580-582; quoted in *Science and Creation*, 40.

[68] Jaki, *Science and Creation*, 40.

[69] Knox, Wisdom 11:20.

[70] Jaki, *Science and Creation*, 41.

The decimal system and notation developed in ancient India between the fourth and seventh centuries represents "the most noteworthy single contribution of ancient India to science and its importance cannot be overstated."[71] Without the decimal system in the Late Middle Ages, the more cumbersome Roman numeral system would have been used and it would have delayed the birth of science in the seventeenth century Christian West.[72] The ancient Indians also built houses out of brick and constructed drainage facilities. There is evidence that they used copper and bronze and made glass.

Advanced technological skill dating back to the third century B.C. and "still unexplained today" is evidenced in the non-rusting pillars erected by King Ashoka during his reign. They certainly remain a monument to progress in metallurgy, stone-cutting, and transportation engineering.[73] The world-famous monuments of the Iron Pillar of Delhi and Sultanganj and copper colossus of Buddha also provide evidence of advanced metallurgy.[74] There was also a lively interest in industrial arts from the third century B.C. in the writing of Kantilya's *Arthasastra*, which articulates the business of government and legislation, the construction of ships, buildings, and roads, and the development of husbandry, agriculture, land surveying, mining, and medicine.[75]

Jaki noted that "practicality, craftsmanship, and organizational talent do not, however, qualify as science."[76] There was no theoretical generalization leading to the

---

[71] Jaki, *Science and Creation*, 13-14.

[72] Jaki, *Science and Creation*, 13-14.

[73] Jaki, *Savior of Science*, 27; Jaki, *Science and Creation*, 14; *Indian History: With Objective Questions and Historical Map*, A-437.

[74] Jaki, *Science and Creation*, 14.

[75] Jaki, *Science and Creation*, 14; Kantilya, Arthasastra (250 B.C.) online at Fordham University, at http://www.fordham.edu/halsall/india/kautilya1.asp.

[76] Jaki, *Science and Creation*, 14.

formulation of physical laws and systems of laws. The claim that science originated in India is also difficult for anyone to make because there are so many doubts about historical sources. This lack of chronology was noted by Needham, a leading historian of not just Chinese, but all Oriental science. He warned that reliable dating of ancient records was necessary for objective analysis of history, and he admitted "the extreme uncertainties in the dating of the most important texts and even of actual objects which have survived" among what is known of Indian history.[77]

Much debate still wages over Needham's infamous question regarding "the failure of China and India to give rise to distinctively modern science while being ahead of Europe for fourteen previous centuries."[78] While acknowledging that the Scientific Revolution was "part of a European miracle," Indian scholars have offered explanations that perhaps India experienced a "mathematical revolution" called "computational positivism" instead. The Indian approach, according to this theory, showed a "deep and studied distrust of axioms and physical models," while Europe "achieved unreasonably and unexpectedly spectacular successes in science."[79]

Other theories have suggested the ability to grasp logical contradictions and "contempt for mundane reality" as a cause for the lack of science in India, while others suggest the cultural stability of agricultural societies with no new

---

[77] Quoted in *Science and Creation*, 22; S. L. Hora, "History of Science and Technology in India and South-East Asia," *Nature*, 168 (1951), 64-65.

[78] Joseph Needham, *Science and the Crossroads*, Papers presented to the 2nd International Congress of the History of Science and Technology (London: 1931), Foreword.

[79] Roddam Narasimha, "The Indian half of Needham's question: some thoughts on axioms, models, algorithms, and computational positivism," *Interdisciplinary Science Reviews*, Vol. 28, No. 1 (2003), 1-13.

challenges to create new knowledge to solve problems.[80] Yet others have alleged that Indian culture was "otherworldly" and perceived the physical world as an illusion and the liberation of the soul more important than the study of the external world.[81] Still others have claimed that since there was no conflict between religion and science in India, and since atheists were not persecuted, this lack of tension is at fault for complacency about science in India; while others have faulted instead the colonialism of the West, and speculated that the "European miracle" would not have happened if it were not for the mathematical contributions of India.[82]

Roddam Narasimha, Indian aerospace scientist and Director of the National Institute of Advanced Studies from 1997 to 2004, where he pursued his interests in the history of science and technology and the philosophy underlying Indic rationalism, referred to the Scientific Revolution as part of a "European miracle" triggered by developments in China and India just as Needham also labeled it. He wrote that the "long Dark Ages of Europe were broken with the help of technical and mathematical inventions imported from the East."[83] This reference was in response to Needham's question regarding the failure of China and India to give rise to modern science while being ahead of Europe for fourteen centuries prior. He called the "birth of modern science" a European rather than a *scientific* miracle because technologies from China and India triggered it, allowing Europe to escape the Dark Ages, by which he meant the "period immediately preceding the birth of modern science" during the two centuries 1500–1700.[84]

[80] D. P. Agrawal and Lok Vigyan Kendra, "The Needham Question: Some Answers," *Indian Science* at http://www.indianscience.org/essays/2-%20NEEDHAMQuestion-DPSameer-edit.pdf.

[81] D. P. Agrawal and Lok Vigyan Kendra, 3-4.

[82] D. P. Agrawal and Lok Vigyan Kendra, 8-13.

[83] Narasimha, 3.

[84] Narasimha, 3.

So, there has been a lot of speculation about why science was not born in India, but the fact remains, it was not. It is true that European science benefited from the technical and mathematical inventions of the East, but it seems impossible to conclude that these inventions were responsible for the birth of modern science, i.e. the "mathematization of science."[85] Modern science is the application, as said, of quantities to natural processes. Both the East and the West had access to nature, but one culture applied mathematics and experimentation to seek an understanding of it and the other cultures did not. If mathematical inventions were responsible for a miracle, then why didn't the culture that invented them experience the so-called miracle first? It seems unsatisfactory to claim that mathematics and technology triggered the birth of modern science in Europe when the real difference in the cultures was a mindset, a psychology, a fundamental way of viewing the universe and existence. Narasimha's argument actually gives support to Jaki's, that science was stillborn in other cultures and born from a Christian mindset.

The Hindus of old also had an animistic view of existence.[86] The doctrine of the Atman represented, and still does represent, a perception of an eternal unity which underlies the phenomenon of nature called the Brahman.[87] Atman is the Indian expression for "first principle," that the individual self of man is found by laying hold of this ultimate self of the universe, the ultimate essences of all things.[88] The

---

[85] Narasimha, 4.

[86] Jaki, *Savior of Science*, 26.

[87] Paul Deussen, *The Philosophy of the Upanishads*, translated by Rev. A. S. Geden, M.A. (Edinburgh: T. & T. Clark, 1906, 1908), 85.

[88] Deussen, *The Philosophy of the Upanishads*, 86.

Indians viewed the universe as an organism, an eternal Pantheistic Being, as did the Egyptians and Chinese.[89]

Indian writings defined the Brahman as a deep sleeper whose vital breath remains dormant, but issues forth on waking, and with his breaths all worlds, gods, and living creatures also awake and are called collectively the Atman.[90] The *pranâs* (speech, eye, ear, touch) proceeded from the Atman.[91] He was the "Soul of the Universe" which bred himself. His mouth, nostrils, eyes, and ears became distinct of his own doing.[92] His skin and hair became the plants and trees, and his heart the moon. His semen became water and his navel exuded corruption.[93] This world-soul is understood as an endless cycle of births and decays with no starting or ending points. Jaki compares it to an eternal "cosmic treadmill."[94]

The Kaliyuga from Indian scripture measured four cycles of human history which were taken to be four ages of the demons characterized by ignorance, poverty, and disease.[95] They should have ended around 300 B.C., but when a golden age did not come, the age of the yugas was recalculated to be 360 years so that, as Jaki interpreted it, their "credibility might be saved."[96] The longer scale meant a never-ending resignation to the age of evil.

Jaki also noted that, in stark contrast to the ancient skill in metallurgy and construction, it was reported by the World Bank in the year A.D. 2000 that only around forty percent of the 825,000 villages in India possessed paved roads to access

---

[89] Jaki, *Science and Creation*, 6.

[90] Deussen, *The Philosophy of the Upanishads*, 87.

[91] Deussen, *The Philosophy of the Upanishads*, 88.

[92] Jaki, *Science and Creation*, 6.

[93] Jaki, *Science and Creation*, 7.

[94] Jaki, *Science and Creation*, 7, Jaki, *Savior of Science*, 26.

[95] Jaki, *Savior of Science*, 27.

[96] Jaki, *Savior of Science*, 27.

essential services.[97] For some reason, technology did not continue even though talent and social stability were not lacking and decimal counting, "possibly the greatest scientific discovery ever made," was invented in ancient India.[98]

As Jaki also showed in other cultures, where there was a pervading resignation to the "cosmic treadmill" or the eternal rebirth of the universe, there was no motivation to try to escape from it. Referencing the *Hymns of the Rig-Veda*, the *Hymns of the Atharva-Veda*, and *Thirteen Principle Upanishads* in *Science and Creation* and *The Savior of Science*, Jaki points out the prose of the eternal cosmic cycle.[99] The Upanishads form the core of Indian philosophy and spiritual teaching, were composed between 800–500 B.C., and are still in use today.[100] From the *Svetasvatara Upanishad* in the First Prapathaka, in the last few lines this resignation is evident:

> In such a world as this what have I to do with the enjoyment of desires? Yea, Even if one were fed therewith to the full, he must still return to earth again and again. Wilt thou therefore deign to deliver me? I am here in this world as a frog in a well without water. Oh adorable one, thou art our refuge, thou art our refuge.[101]

Jaki also quoted a twentieth-century explanation from Gandhi in 1938 about the absolute superiority of life with no

---

[97] "Rural Roads: A Lifeline for Villages in India," The World Bank (New Delhi) World Bank website, http://web.worldbank.org/.

[98] Jaki, *Savior of Science*, 28.

[99] Jaki, *Science and Creation*, 7, 21-22; Jaki, *Savior of Science*, 28.

[100] *Indian History with Objective Questions and Historical Map*, A-117.

[101] Maitri Upanishad, *Sanskrit Text with English Translation*, edited and translated by E. B. Cowell (London: Asiatic Society of Bengal, 1870), 244.

technology.[102] In this statement Gandhi echoed the philosophy and teaching of the Upanishad of ancient times:

> I believe that the civilization that India evolved is not to be beaten in the world. . . . India remains immovable and that is her glory. . . . Our ancestors dissuaded us from luxuries and pleasure. We have managed with the same kind of plow as existed thousands of years ago. ... We have had no system of life-corroding competition. . . . It was not that we did not know how to invent machinery, but our forefathers knew that, if we set our hearts after such things, we would become slaves and lose our moral fibre. They, therefore, after due deliberation decided that we should only do what we could with our hands and feet. . . . They were, therefore, satisfied with small villages. . . . They held the sovereigns of the earth to be inferior to the Rishis and the Fakirs. A nation with a constitution like this is fitter to teach others than to learn from others.[103]

This is not to imply that there is nothing beautiful in labor and toil for the needs of life, but the implications of these ancient teachings for Indian science are also indisputable. There was a psychology not conducive to the birth of modern science even though the skill was apparent long ago. "Science," Jaki wrote, "cannot arise, let alone gain sustained momentum, without an articulated longing for truth which in turn presupposes a confident approach to reality."[104]

---

[102] Jaki, *Savior of Science*, 29.

[103] M. K. Gandhi, *A Dialogue between an Editor and a Reader*, Hind Swaraj, or Indian Home Rule (1938) M. K. Gandhi website, http://www.mkgandhi.org/swarajya/, "What is True Civilization;" *Savior of Science*, 29.

[104] Jaki, *Science and Creation*, 19.

Babylon

Jaki next, in *The Savior of Science*, offered the history of science among cultures that communicated and developed in succession—Babylon, Greece, and Arabia. Knowledge was transmitted to the Sumerians from the Egyptians and then on to Babylonians, Assyrians, and Persians (c. 2900 B.C.— mid 7th century A.D.). From there knowledge was transmitted to the Greeks and then to the Arabs, and this history is recorded in greater detail. Again, just as in other ancient cultures, there was obvious skill.

Jaki devoted the first few pages of the chapter "The Omen of Ziggurats" in *Science and Creation* to those massive structures built by the Sumerians and Babylonians as an example of their technological ability. "The planning, building, and decorating of the ziggurats," Jaki wrote, "implied craftsmanship and practical geometry and is application on a grand scale, especially if one considers the temple complex and city surrounding the ziggurat."[105] Towns were planned around these temples, with defense walls, palaces, and quarters of the cities.[106] The temples were made of mud bricks laid in a herring-bone pattern with mud mortar overlaid with bitumen or lime plaster, which could be smoothed and polished to a high-quality finish, to waterproof them.[107] According to Herodotus, an ancient Greek historian (c. 484–425 B.C.), the ziggurat of Babylon was exceeded by no other city: "[Babylon] lies in a great plain, and is in shape a square, each side fifteen miles in length; thus sixty miles make the complete circuit of the city. Such is the size of the city of

---

[105] Jaki, *Science and Creation*, 85.

[106] Harriet Crawford, *Sumer and the Sumerians*, Second Edition (Cambridge, UK: Cambridge University Press, 2004), 60.

[107] Crawford, *Sumer and the Sumerians*, 67.

Babylon; and it was planned like no other city of which we know."[108]

Jaki credited the Babylonian discovery of "mathematical puzzles equivalent to second-degree equations, lists of hundreds of plants and chemical compounds, together with their astonishingly accurate medicinal properties, and even longer lists of planetary positions" as extraordinary items of learning.[109] The lists of planetary positions were the factual proof that Hipparchus, a second-century B.C. Greek astronomer and mathematician who discovered the precession of the equinoxes, relied on Babylonian astronomical data to reach his conclusions, another "one of the greatest scientific discoveries of all times."[110] The Babylonians had the skills necessary to apply mathematics to nature, but they did not take this step.

It may seem contradictory for Jaki to have claimed that science was "born" of Christianity and "stillborn" in other cultures while he also credits those cultures with great scientific discoveries. This point is often missed by other historians, so it is useful to pause here to point out something in Jaki's use of the word "stillbirth" of science. He acknowledged cultural wombs that were capable of developing science even to the point of viability as a sustained discipline. His choice of the word "birth" was to show that the final step from isolated dependence to universal independence was not taken in any culture before the Scientific Revolution in the Middle Ages.

Jaki devoted more analysis of Babylonian science and the reasons why it was a "stillbirth" in *Science and Creation*. Historians have also argued that "all subsequent varieties of

---

108 Herodotus, *The Histories*, with an English translation by A.D. Godley (Cambridge, UK: Harvard University Press, 1920), Book 1, Chapter 178, section 3.

109 Jaki, *Savior of Science*, 38.

110 Jaki, *Savior of Science*, 38.

scientific astronomy, in the Hellenistic world, in India, in Islam, and in the West–if not indeed all subsequent endeavors in the exact sciences–depend upon Babylonian astronomy in decisive and fundamental ways."[111] Jaki would not have agreed with this argument, and the reason has to do with how he carefully defined "exact science" and how he considered the theological implications of ancient cultures as well. All subsequent varieties of science may have depended in some way on Babylonian astronomy, but not in decisive and fundamental ways.

The Babylonians may have mathematically modeled astronomical appearances, but in their cosmology it is evident that they believed a very different reality existed *behind* the appearances.[112] The *Enuma elish* was a portrayal of personified forces engaged in bloody battles; the mother goddess, Tiamat, is dismembered to form the sky, earth, waters, and air.[113] Jaki explained, "Such a cosmogony was certainly not a pointer toward that kind of understanding of the cosmos which amounts to science."[114]

The Babylonians observed celestial phenomenon as a service to a religious worldview steeped in magic, and the calculations were abstracted from physical objects. Jaki held that it was absolutely necessary for a true science, the "*quantitative* study of the quantitative aspects of physical *objects* in motion," for calculations not to be abstracted from objects. A viable birth of science could not have been made in such an environment where the mathematical formality was cut off from the physical reality. The failure was neither geophysical nor socio-economical, but rather an intellectual

---

[111] A. Aaboe, "Scientific Astronomy in Antiquity," *Philosophical Transactions of the Royal Society of London. Series A, Mathematical and Physical Sciences*, A. 276 (1974), 21-42.

[112] Jaki, *Science and Creation*, 89.

[113] Jaki, *Savior of Science*, 39.

[114] Jaki, *Savior of Science*, 39.

inertia that prevented a systematic investigation of the world and its lawfulness. There was no confidence in the reasonability of such an enterprise under the belief that people were part of a huge, animistic, cosmic struggle between chaos and order.[115]

Greece

Like other great civilizations, the contributions and skill of the ancient Greeks cannot be dismissed. Probably more has been written about Greek intellectual history than any other ancient culture. Many scholars have credited ancient Greece with the invention of science, and Jaki held that they came closer to a birth of science than any other culture. There is a long list of scholars who left behind writings that inspire intellectual endeavors to this day. Here is a brief list of some of them.

Thales of Miletus (c. 620–c. 546 B.C.) was a geometer and astronomer influenced by the Babylonians and Egyptians. He developed ideas about abstract geometry such as the idea that the diameter of a circle bisects the circle and that base angles of isosceles triangles are equal.[116] He is popularly credited as the "first scientist" even by those who admit that in the strict sense the recipient of this title is unknown because "science as a reliable method to knowledge, involving observation, hypothesis, experiment, and critique, would evolve."[117] Centuries later, Aristotle of Stagira, who will be discussed

---

[115] Jaki, *Science and Creation*, 99.

[116] Dirk L. Couprie, "How Thales Was Able to 'Predict' a Solar Eclipse Without the Help of Alleged Mesopotamian Wisdom," *Early Science and Medicine* Vol. 9, No. 4 (2004), 321-337.

[117] Robert McHenry, "Thales of Miletus: The First Scientist, the First Philosopher," *Encyclopedia Britannica Blog*, at Encyclopedia Britannica, Inc., http://www.britannica.com/blogs/2010/04/thales-of-miletus-hero/.

shortly, called Thales the "founder" of the type of philosophy that investigates the nature of matter and original causes.[118] Thales was the founder of the school of thought known as Ionian physics. He conceived of the world of nature as an organism, an animal, within which were lesser organisms.[119] The earth was, according to the Ionians, one such organism in the greater organism and it served its own purpose.

Anaximander of Miletus (c. 611–c. 547 B.C.) described the origin of all things as the "Boundless" or the "Unlimited" principle and was a speculative astronomer who wrote about celestial bodies and why the Earth does not fall.[120] An Ionian following in the thought of Thales, Anaximander considered time and space as a matrix of birth to successive worlds.[121] He thought that innumerable worlds arose in this boundless medium like bubbles and that the earth is but one of those bubbles. Ionian physics presupposed that all natural things were made of a single substance. Anaximander proposed that the cosmos was less like a god-like organism and more like a divine substance.[122]

Pythagóras of Sámios (c. 570–c. 490 B.C.) was a famous mathematician for the theorem named after him, and although the most authoritative history of early Greek geometry assigns him no role in geometry at all, his discoveries were significant nonetheless.[123]

Leucippus (fifth century B.C.) is considered the founder of Atomism, along with Democritus (c. 460–c. 370 B.C.) The

---

[118] Aristotle, *Metaphysics*, Book 1, Part 3.

[119] Robin George Collingwood, *The Idea of Nature* (Oxford: Oxford University Press, 1945), 30.

[120] Dirk L. Couprie, "Anaximander," *Internet Encyclopedia of Philosophy* (2001/2005) at http://www.iep.utm.edu/anaximan/.

[121] Collingwood, 33.

[122] Collingwood, 35-36.

[123] Carl Huffman, "Pythagoras", *The Stanford Encyclopedia of Philosophy* (Fall 2011 Edition), Edward N. Zalta (ed.), at http://plato.stanford.edu/archives/fall2011/entries/pythagoras/.

theory of Atomism held that there could be no motion without voids and that invisible and indivisible particles moved in the empty space. Democritus called these particles ἄτομος or *atomos*, a term which is still used today. The theory was a philosophical one, not based on observation or experiment, to explain how there might be change without something coming to be out of nothing.[124]

Hippocrates of Cos (c. 450–c. 380 B.C.), among others, is credited with providing detailed medical observations that made it possible to diagnose and treat illness, along with a code of ethics that still has influence today.[125] The Greeks borrowed from the Babylonian intellectual treasures, but they also developed their own system of geometry, without which the Babylonian data would have been unbeneficial.[126] Modern geometers are still unable to reconstruct the demonstrations behind some propositions in the Fourteen Books of Euclid.[127] Euclid of Alexandria (c. 325–265 B.C.) built a logical and rigorous geometry with a solid foundation, and it was a primary source of geometric reasoning and methods that went practically unchanged for more than two thousand years.[128]

---

[124] Sylvia Berryman, "Leucippus," *The Stanford Encyclopedia of Philosophy* (Fall 2010 Edition), Edward N. Zalta (ed.), at http://plato.stanford.edu/archives/fall2010/entries/leucippus/; Sylvia Berryman, "Democritus," *The Stanford Encyclopedia of Philosophy* (Fall 2010 Edition), Edward N. Zalta (ed.), at http://plato.stanford.edu/archives/fall2010/entries/democritus/.

[125] Michael Boylan, "Hippocrates," *Internet Encyclopedia of Philosophy* (2002/2005) at http://www.iep.utm.edu/hippocra/.

[126] Jaki, *Savior of Science*, 39.

[127] Jaki, *Savior of Science*, 40; for example see, Edward Grant, *A Source Book in Medieval Science* (Cambridge, MA: Harvard University Press, 1974), 159, Footnote 7 about irrational ratios.

[128] Christian Marinus Taisbak, "Euclid," Encyclopedia Britannica (2013) at http://www.britannica.com/EBchecked/topic/194880/Euclid.

There was, of course, the great Aristotle of Stagira (384–322 B.C.), the major Greek philosopher and student of the great Plato, the teacher who founded the Lyceum in Athens. Aristotle wrote stupendous volumes on logic, politics, biology, taxonomy, physics, and cosmology.[129] The most mature form of science achieved during Hellenic times in the biological sciences was that of Aristotle's. He turned zoology into a scientific discipline in his *History of Animals* and laid the foundations for comparative anatomy in his *On the Parts of Animals*. His *On the Generation of Animals* remained, until modern times, the authority on embryology.[130]

Jaki wrote that the "extraordinary feats of Aristotle in biology were in a sense responsible for his failure in physics."[131] According to Jaki, Aristotle's *On the Heavens* "set the fate and fortune of science, or rather tragic misfortunes, for seventeen hundred years" because a serious error was made and went unnoticed in Aristotle's continuous resort to biological simile.[132] In holding the belief that all things had a soul and therefore sought the final cause for which they were best suited (i.e. rocks desire to fall to the ground), for animals as well as for objects, a purpose was assumed for processes and phenomena of every kind.

Aristotle asserted that if two bodies were dropped from the same height at the same time, the one with twice the weight of the other one would fall twice as fast because it had twice the nature and twice the desire to do so.[133] Even though simple observation would prove that false, the hold on the

---

[129] Christopher Shields, "Aristotle," *The Stanford Encyclopedia of Philosophy* (2013), Edward N. Zalta (ed.), at http://plato.stanford.edu/archives/fall2013/entries/aristotle/.

[130] Jaki, *Science and Creation*, 104.

[131] Jaki, *Science and Creation*, 104.

[132] Jaki, *Savior of Science*, 40; Jaki, *Science and Creation*, 105.

[133] Aristotle, *On the Heavens*, Book 1, Part 6, third paragraph, "…"if one weight is twice another, it will take half as long over a given movement."

mind of the Greeks of this animistic orthodoxy would not allow it. The Greeks thought of motion as a function of the magnitude, a "striving," in nature for objects living and non-living. Aristotle dismissed the idea of unresisted motion as unreal or over-abstract.[134] This orthodoxy caused even a genius like Aristotle to be so wrong about the free fall of objects. It is perplexing that no one noticed this falsehood in daily life, not just among the ancient Greeks but, as will be discussed later, also among those who followed Aristotle's orthodoxy into the thirteenth and fourteenth centuries.

These views were—as has been noted in the Egyptian, Chinese, Indian, and Babylonian cultures already—the result of pantheism. The Greeks were steeped in the perspective of eternal cycles of birth-life-death-rebirth for all things, a theme common to all the great religions and cultures that experienced a stillbirth of science. In keeping with the mindset of Babylonian and Egyptian cultures, the Greeks also put a strong emphasis on an eternal, cyclic universe and on the comparison of the cosmos to animals. Even the ones with a belief in a monotheistic deity believed that deity *was* the universe and that all existence was a cyclic "cosmic treadmill."

With Plato, Socrates, and Aristotle, the sublunary world was like a huge animal breathing, growing, and decaying in cycles of birth, death, and rebirth for eternity.[135] The basis of this belief was that fundamentally all existence was viewed as cyclical and the cosmos either was a god-organism or a god-substance obeying unpredictable laws of its own volition and from which finite substances came into and out of existence. In subsequent centuries, even as new ideas about the nature of the cosmos were explored, the fundamental cyclical presumption remained. Philolaus, Alcmeon, Archytas, and Oenipodus were Pythagoreans who promoted a cyclical

---

[134] Stephen Toulmin, *Foresight and Understanding: An Enquiry into the Aims of Science* (New York, NY: Harper & Row, 1961), 51.

[135] Jaki, *Science and Creation*, 105.

constitution of the universe.[136] This belief was not irrational since human experience is based on cycles in life, in nature, and in the heavens.

Plato, in his many dialogues, told of a cosmic process that alternated between two phases, one of divine laws with a golden age and one of chaos and destruction. In *Republic*, Plato explains how he saw everything, including human and social phenomenon and the periodicity of human societies, under the organic cosmic law of cycles.[137]

> A city which is thus constituted can hardly be shaken; but, seeing that everything which Hard in truth it is for a state thus constituted to be shaken and disturbed; but since for everything that has come into being destruction is appointed, not even such a fabric as this will abide for all time, but it shall surely be dissolved, and this is the manner of its dissolution. Not only for plants that grow from the earth but also for animals that live upon it there is a cycle of bearing and barrenness for soul and body as often as the revolutions of their orbs come full circle, in brief courses for the short-lived and oppositely for the opposite; but the laws of prosperous birth or infertility for your race, the men you have bred to be your rulers will not for all their wisdom ascertain by reasoning combined with sensation, but they will escape them, and there will be a time when they will beget children out of season.[138]

This cycling between chaos and divine order came to be called the Great Year (sometimes Perfect Year), the time when all the stars and constellations of the sky came back to

---

[136] Jaki, *Science and Creation*, 106.

[137] Jaki, *Science and Creation*, 110.

[138] Plato, *The Republic*, translated by Paul Shorey. Volume II (Cambridge, MA: Cambridge University press, 1942), 245-247.

the position they were in a golden age of perfection, the period of one complete cycle of the equinoxes, although that date was figured differently by different philosophers and thus was ambiguous.[139] However, the general point is that the notion of a cyclical eternity was prevalent and persistent in Greek thought. In *Timaeus*, Plato describes this Great Year:

> Thus arose day and night, which are the periods of the most intelligent nature; a month is created by the revolution of the moon, a year by that of the sun. Other periods of wonderful length and complexity are not observed by men in general; there is moreover a cycle or perfect year at the completion of which they all meet and coincide . . . To this end the stars came into being, that the created heaven might imitate the eternal nature.[140]

The gods, according to Plato and the Greeks, were themselves made in the form of a circle, the "most perfect figure and the figure of the universe." According to Aristotle, time itself was, therefore, a circle.[141] If time is a circle and the cosmos eternal within this circle, emanating from the pantheistic God, the nature of the gods, un-aging, un-alterable, and un-modified, then all change, including human knowledge, is cyclical too. For the most brilliant scholar or the least accomplished servant, the Greeks believed the same thoughts are recurring over and over again, and Aristotle held that this was, in fact, what man experienced:

---

[139] James Adams, *The Republic of Plato: Edited with Critical Notes, Commentary, and Appendices*. Second Edition. Volume II. Books VI-X and Indexes (Cambridge, MA: Cambridge University Press, 1963), 303-304.

[140] Plato, *Republic*, 245-247.

[141] Jaki, *Science and Creation*, 130.

The mere evidence of the senses is enough to convince us of this, at least with human certainty. For in the whole range of time past, so far as our inherited records reach, no change appears to have taken place either in the whole scheme of the outermost heaven or in any of its proper parts. The common name, too, which has been handed down from our distant ancestors even to our own day, seems to show that they conceived of it in the fashion which we have been expressing. The same ideas, one must believe, recur in men's minds not once or twice but again and again.[142]

Jaki described the psychological impact of this premise as complacency. Such a belief hardly inspired an intellectual curiosity or confidence to learn and dominate the physical laws of nature, even if one felt that he was living in a golden age.

Clearly, if one is consciously merged into the treadmill of eternal recurrences, only two choices remain. One is that of hopelessness, the feeling that one is at the bottom. The other is complacency, the illusion that one is and remains on top, at least in the sense that the irreversible decline will begin to be felt only by one's distant progeny. Both attitudes cry out for salvation, although the second may be the less receptive to it.[143]

The psychological impact is inherently tied to the cyclic worldview and the Great Year. Even in times of great progress, such as the Greek civilizations experienced, there would have been a resignation that the human cannot escape whatever fate pantheism and animism held for him. In eras of

---

[142] Aristotle, *On the Heavens*, Book I, Part 3.
[143] Jaki, *Savior of Science*, 44.

despair, there would have been a resignation to wait it out, even beyond one's lifetime.

Arabia

The last culture to be examined is that of the Muslims. Although theirs was a monotheistic view, it was not a Christological or Trinitarian view, which left it vulnerable to a monotheism that approached pantheism. What happened in the Muslim world seems to be the result of a mixture of mindsets. The Arabian philosophers adopted the works of the Greeks, along with the organismic, eternal cosmic treadmill worldview. This meant that the philosophers' worldview was in conflict with the Muslim religion since the Koran taught that God the Creator created the world and held it in existence. The stillbirth of Muslim science could be credited with a *separation* of science and religion that ought to have been reconciled, a point that would no doubt surprise many people today.

As Athens and Rome lost cultural significance around the early seventh century A.D., there was less communication between the two. Greek scholars moved toward the East and organized at Jundishapur in Southwest Persia. In 641 A.D., when Persia was conquered by the Muslims, the Middle East and North Africa came under one rule.[144] By 711 A.D., the Arabs took Spain and twenty-one years later they stormed France. One hundred years after Muhammed's death, a political unification of land that spanned three continents emerged. As the new religion codified in the Koran was imposed, a giant empire formed "steeped in the conviction that everything in life and in the cosmos depended on the sovereign will of a personal God, the Creator and Lord of all."[145]

---

[144] Jaki, *Science and Creation*, 192.
[145] Jaki, *Science and Creation*, 193.

The continual study of the Koran inspired intellectual curiosity among faithful Muslims, as did the meticulous scholarship of the Greek philosophical and scientific body of knowledge.[146] So serious was the promotion of knowledge that "Houses of Wisdom" were erected, notably in Baghdad (813–833), Cairo (966), and Cordova (961–976).[147] Cordova amassed over 300,000 volumes for the library and immediately attracted scholars from the Christian West, who were welcomed with hospitality.[148]

A paper mill, learned from the Chinese art of paper-making, was constructed in Baghdad in 794, and extensive translation and reproduction of Greek literature flourished. The works of Galen, who was considered second only to Hippocrates in the medical hagiography of the Western World, were translated, some 130 of them, and dominated medical practice in the medieval East and the West well into the Renaissance.[149] The greatest figure of Arab medicine was produced from this school, al Razi (865–925), the author of *A Treatise on the Small-Pox and Measles*. His work has been reprinted more than forty times in the last four hundred years.[150] Islamic medicine in general was outstanding, a field in which Islamic science demonstrated its most sustained vitality. The Muslims had a realistic sense for facts of observation.

The Islamic ophthalmologist, Ibn-Rushd (1126–1198), otherwise known as Averroes, provides a "priceless insight" into the ultimate failure of Islamic science.[151] He was a resolute advocate and student of Aristotle's philosophy and

---

[146] Jaki, *Savior of Science*, 44

[147] Jaki, *Science and Creation*, 194.

[148] Jaki, *Science and Creation*, 194.

[149] Prioresch Plinio, *A History of Medicine: Roman Medicine* (Omaha, NE: Horatius Press, 1998) 315; Jaki, *Science and Creation*, 194.

[150] Jaki, *Science and Creation*, 194.

[151] Jaki, *Science and Creation*, 195.

science, and as such broke new grounds with ophthalmology. The practice of medicine could flourish under Aristotelian teaching because it did not require any questioning of Aristotle's view of the *physics*.

Likewise Ibn-Sina (980–1037), also known as Avicenna, the famed philosopher provides the same insight.[152] His textbook served as the standard in Arab medical teaching, a fine collection of observation and systematic pathology. Muslim science made notable contributions in areas that had nothing to do with *physical* laws. When it came to a study of physical laws of the world, there was a certain inertia owed to the unwillingness to question the Aristotelian animistic worldview, which is why the study of biology advanced but without an underlying increase in the understanding of the physical world.[153]

This lack of understanding of physics is evidenced by Arab alchemy, which came to stand for the study of materials and compounds. This field of investigation was a combination of "mystical and astrological proclivities," fundamentally the result of mixing the organismic, eternal cycles of pantheism with the belief that a Creator created the universe.[154] It was an attempt to reconcile the conflicting views of Aristotelian philosophy and Muslim theology.

The same paradox occurred in astrology. The astrologers, working with assumptions in conflict with their religion, gave credence to the pagan doctrine of the Great Year, even to the point of believing it could predict the succession of rulers, religions, reigns, and physical catastrophes.[155] Yet devout Muslims could not accept these ideas that were in conflict with Muslim orthodoxy, which revealed that the universe had

---

[152] Jaki, *Science and Creation*, 195.
[153] Jaki, *Science and Creation*, 197.
[154] Jaki, *Science and Creation*, 197.
[155] Jaki, *Science and Creation*, 198.

an absolute beginning with creation.[156] As attempts were
made to reconcile these beliefs, something ambiguous
resulted, as evidenced in the writing of al-Biruni, a Muslim
who refuted the contradictions among scholars and religious
men in his famous work *The Chronology of Ancient Nations*:

> It is quite possible that the (celestial) bodies were
> scattered, not united at the time when the Creator
> designed and created them, they having these motions, by
> which–as calculation shows–they must meet each other in
> one point in such a time (as above mentioned). It would
> be the same, as if we, e.g. supposed a circle, in different
> separate places of which we put living beings, of whom
> some move fast, others slowly, each of them, however,
> being carried on in equal motions–of its peculiar sort of
> motion–in equal times; further, suppose that we knew
> their distances and places at a certain time, and the
> measure of the distance over which each of them travels
> in one Nychthemoreon.[157]

He goes on in the work to give credit to the mathematical
computations of the cycles to explain the appearances, an
incongruity between mathematics and reality and a failure to
go beyond the Aristotelian and Neoplatonian positions
regarding the physical world.[158] As far as the Muslim scholars
advanced, they still did not provide the psychology that could
give birth to modern science because they did not effectively
refute the pantheism of the Greek scientific *corpus* (body).

---

[156] Jaki, *Science and Creation*, 199.

[157] Athar-ul-Bakiya of Albiruni, *The Chronology of Ancient Nations*:
An English Version of the Arabic Text "Vestiges of the Past,"
translated by D. Edward Sachau (London: W. H. Allen & Co.,
1879), 30; quoted in *Science and Creation*, 199.

[158] Jaki, *Science and Creation*, 200.

The Biblical Womb

The Biblical Womb

"Who was it measured out the waters in his open hand, heaven balanced on his palm, earth's mass poised on three of his fingers?"[159]

In *Science and Creation*, Jaki included the ancient Hebrews in this history of the development of science, people who are not usually considered since their culture centered on religious law. Jaki highlighted that in this culture there was a literary codification of the concept of a Creator and of a creation out of nothing, the point of the book of Genesis. This concept was a radical break from Egyptian, Babylonian, and Greek thought, but it was the same codification as the Koran. There are also detailed references in the books of the prophets and the psalms to the faithfulness of the regular and permanent structure and function of nature, offered repeatedly as the basis for believing in the trustworthiness of God. It may seem trivial to note that the cultures of the Old Testament viewed the universe as created and ordered since most scholars focus exegesis on the study of salvation history in the Old Testament, but Jaki highlighted a significant pre-scientific historical message, too.[160]

The Prophets

When the Israelites in Babylian captivity hoped for the restoration of Jerusalem, the Word of the Lord came to Jeremiah to remind the people that it is God who orders the day and night and God who promised heirs to David's throne with a posterity "countless as the stars of the heaven, measureless as the sea-sand."[161] The Israelites knew they must

---

159 Knox, Isaiah 40:12
160 Jaki, *Savior of Science*, 61.
161 Knox, Jeremiah 33:20-21; Jaki, *Savior of Science*, 62.

trust the faithfulness of God because they knew that the humans are not the ones who order the day and night.

> A message from the Lord, from him, the God of hosts, the same who brightens day with the sun's rays, night with the ordered service of moon and star, who can stir up the sea and set its waves a-roaring. All these laws of mine will fail me, he says, before the line of Israel fails me; a people it must remain until the end of time. You have the Lord's word for it; When you can measure heaven above, he tells you, and search the foundations of earth below, then I will cast away the whole line of Israel, for all its ill deserving.[162]

This is an established prophetic tradition in the Old Testament cultures. The law of God extends to all things moral, societal, and natural. The nations are told to submit to God's will and obey His commands. God's unchallenged power is often mentioned in Isaiah. Consider the drastic difference in the naturalistic mindset of Isaiah compared to the writings of ancient Egypt, India, China, Babylon, and Greece. God is the Creator, not the universe. Isaiah points to the order and measure of physical objects, i.e. what Jaki defined as "exact science," as contributing proof of God's *omniscience*, literally translated to mean "having *all* knowledge."

> Who was it measured out the waters in his open hand, heaven balanced on his palm, earth's mass poised on three of his fingers? Who tried yonder mountains in the scale, weighed out the hills?

> No aid, then, had the spirit of the Lord to help him, no counsellor stood by to admonish him. None other was there to lend his skill; guide to point out the way, pilot to

---

[162] Knox, Jeremiah 31:35-37; Jaki, *Savior of Science*, 62.

warn him of danger. What are the nations to him but a drop of water in a bucket, a make-weight on the scales? What are the islands but a handful of dust? His altar-hearth Lebanon itself could not feed, victims could not yield enough for his burnt-sacrifice. All the nations of the world shrink, in his presence, to nothing, emptiness, a very void, beside him. And will you find a likeness for God, set up a form to resemble him? What avails image the metal-worker casts, for goldsmith to line with gold, silversmith plate with silver? What avails yonder wood, hard of fibre, proof against decay; the craftsman's care, that his statue should stand immovable?

What ignorance is this? Has no rumour reached you, no tradition from the beginning of time, that you should not understand earth's origin? There is One sits so high above its orb, those who live on it seem tiny as locusts; One who has spread out the heavens like gossamer, as he were pitching a tent to dwell in. The men who can read mysteries, how he confounds them, the men who judge on earth, what empty things he makes of them! Saplings never truly planted, or laid out, or grounded in the soil, see how they wither at his sudden blast, how the storm-wind carries them away like stubble! What likeness, then, can you find to match me with? asks the Holy One. Lift up your eyes, and look at the heavens; who was it that made them? Who is it that marshals the full muster of their starry host, calling each by its name, not one of them missing from the ranks? Such strength, such vigour, such spirit is his.[163]

The naturalness of the universe, the predictability and order, the power of God as Creator and Lawmaker are all emphasized, indicating a view of the cosmos that was

---

[163] Knox, Isaiah 40:12-26; Jaki, *Savior of Science*, 63.

sustained leading up to and during the birth of modern science. The absolute certainty of the faithfulness of God is invoked to give credibility to the belief that Jerusalem will be rebuilt: "It was I framed the earth, and created man to dwell in it; it was my hands that spread out the heavens, my voice that marshaled the starry host."[164] The Old Testament people saw nothing that happened in nature as vain; even the rain that falls from the sky makes the land fruitful.[165] Yahweh alone, who created nature, can bring nature to an end and final judgment of all.[166] Genesis 1 is much more rational than the *Enuma elish* creation myth of Babylon in which personified forces engaged in bloody battles dismembering the mother goddess, Tiamat, to form the sky, earth, waters, and air.

This mindset, this view of the universe in Genesis 1, permeated the thought of the Israelites, the Jews, and the early Church. The formulations about the universe in the Old Testament would act as the "mind's saving grace" in the birth of science in the European Middle Ages because it implied nothing that could have been perceived by reason, experiment, or observation alone; it is a revelation, and a necessary one for the birth of science.[167]

The Psalms

In the psalms is found a poetic conviction regarding the work of Creation and its relevance to everything man thinks or does.[168] The monotheistic outlook on the world is unmistakable and uncompromising, enthusiastic even. This striking confidence is abundantly evident, and shows the

---

[164] Knox, Isaiah 45:12, 18; Jaki, *Savior of Science*, 63.

[165] Knox, Isaiah 55:10; Jaki, *Savior of Science*, 63.

[166] Knox, Isaiah 23:10-11; 24: 21-23; Jaki, *Savior of Science*, 63.

[167] Jaki, *Savior of Science*, 59.

[168] Jaki, *Science and Creation*, 148.

belief in Creation of the entire cosmos out of nothing as well as a belief in the miraculous Creator who could accomplish the former obviously could produce the latter. Even in the earliest psalms, there is a most confident vision of nature, a precursor of the science to come. This familiar psalm is not usually taken as an indicator of pre-scientific attitudes, but compared to the creation myths of other ancient cultures and the pantheistic worldview they held, this view of a caring Creator of nature stands out.

> The Lord is my shepherd; how can I lack anything?
> He gives me a resting-place where there is green pasture,
> leads me out to the cool water's brink, refreshed and content.
> As in honour pledged, by sure paths he leads me;
> dark be the valley about my path,
> hurt I fear none while he is with me;
> thy rod, thy crook are my comfort.
> Envious my foes watch, while thou dost spread a banquet for me;
> richly thou dost anoint my head with oil, well filled my cup.
> All my life thy loving favour pursues me;
> through the long years the Lord's house shall be my dwelling-place.[169]

The universe of the Old Testament is good, complete, and ordered. The universe is not a creature of unpredictable volition, but the creation of a personal and loving Creator. There is no conflict between reason and revelation, and the order, stability, and predictability of the cycles of the cosmos testify to the faithfulness of God.

---

[169] Knox, Psalm 23.

Give thanks to the Lord for his goodness, his mercy is eternal; give thanks to the God of gods, his mercy is eternal; give thanks to the Lord of lords, his mercy is eternal. Eternal his mercy, who does great deeds as none else can; eternal his mercy, whose wisdom made the heavens; eternal his mercy, who poised earth upon the floods. Eternal his mercy, who made the great luminaries; made the sun to rule by day, his mercy is eternal; made the moon and the stars to rule by night, his mercy is eternal.

Eternal his mercy, who smote the Egyptians by smiting their first-born; eternal his mercy, who delivered Israel from their midst, with constraining power, with his arm raised on high, his mercy is eternal. Eternal the mercy that divided the Red Sea in two, eternal the mercy that led Israel through its waters, eternal the mercy that drowned in the Red Sea Pharaoh and Pharaoh's men. And so he led his people through the wilderness, his mercy is eternal.

Eternal the mercy that smote great kings, eternal the mercy that slew the kings in their pride, Sehon king of the Amorrhites, his mercy is eternal, and Og the king of Basan, his mercy is eternal. Eternal the mercy that marked down their land to be a dwelling-place; a dwelling-place for his servant Israel, his mercy is eternal. Eternal the mercy that remembers us in our affliction, eternal the mercy that rescues us from our enemies, eternal the mercy that gives all living things their food. Give thanks to the God of heaven, his mercy is eternal.[170]

The reason for quoting such a long passage is to show the richness and unity of the outlook of the world and humanity

---

[170] Knox, Psalm 136; Jaki, *Science and Creation*, 148-149.

that derived from Genesis 1. There are many other passages that describe the same. God is not just a dispassionate creator; He is eternally merciful and faithful to His people, and that faithfulness is evidenced in the stability of creation. There is an abundance of such praises in Psalms 35, 80, and 120 of the stability of nature as a work of the Creator. Psalm 73, for example, praises God's hold on creation: "Thine is the day, thine the night; moon and sun are of thy appointment; thou hast fixed all the bounds of earth, madest the summer, madest the cool of the year."[171] Psalms 118 praises God for the stability of the moral law as well as nature: "Lord, the word thou hast spoken stands ever unchanged as heaven. Loyal to his promise, age after age, is he who made the enduring earth."[172] Passages such as these demonstrate the naturalness of order and stability in creation. The enduring process of nature also was taken as proof of the certainty of God's enduring rule: "Ageless as sun or moon he shall endure; kindly as the rain that drops on the meadow grass, as the showers that water the earth. Justice in his days shall thrive, and the blessings of peace; and may those days last till the moon shines no more."[173]

There are many more, and Jaki lists them in both *Science and Creation* and *The Savior of Science*; they all provide evidence that the Bible is the backdrop for the worldview that God created a physical realm that is stable and that this stability gives testimony to the stability of God's moral law and to His faithfulness. It was not just an explanation for nature, but an explanation for all creation.

Wisdom Literature

---

[171] Knox, Psalm 73: 16-17; Jaki, *Savior of Science*, 65.
[172] Knox, Psalm 118: 89-90; Jaki, *Savior of Science*, 65.
[173] Knox, Psalm 71: 5-7; Jaki, *Savior of Science*, 65.

The Wisdom literature is especially substantial evidence of this Old Testament worldview. In the first three chapters of the Book of Proverbs there are three series of instructions about wise behavior, and the starting point is a reference to God's wisdom in the created world, its stability, and the firmness of the heavens and the earth. The praise of wisdom goes for five more chapters and ends with a personification of God's wisdom:

> The Lord made me his when first he went about his work, at the birth of time, before his creation began. Long, long ago, before earth was fashioned, I held my course.

> Already I lay in the womb, when the depths were not yet in being, when no springs of water had yet broken; when I was born, the mountains had not yet sunk on their firm foundations, and there were no hills; not yet had he made the earth, or the rivers, or the solid framework of the world.

> I was there when he built the heavens, when he fenced in the waters with a vault inviolable, when he fixed the sky overhead, and levelled the fountain-springs of the deep.

> I was there when he enclosed the sea within its confines, forbidding the waters to transgress their assigned limits, when he poised the foundations of the world.

> I was at his side, a master-workman, my delight increasing with each day, as I made play before him all the while; made play in this world of dust, with the sons of Adam for my play-fellows.

> Listen to me, then, you that are my sons, that follow, to your happiness, in the paths I show you; listen to the

teaching that will make you wise, instead of turning away from it.[174]

The Old Testament is the story of the unity of cosmic and human history. The Maker of the World is also the Shepherd of His People.[175] The Book of Wisdom was written in Alexandria around the first century before Christ as Jewish thinkers came into contact with Hellenistic learning in Alexandria. There was a cultural refinement between the polytheistic nature worship of the Greeks and the creation *ex nihilo* in the unique worldview of the people of the Covenant.[176] The author of the Book of Wisdom appreciates the knowledge of the Hellenistic culture but views it through the perspective gained by a belief in a Creator of the universe, the source of all wisdom:

> Sure knowledge he has imparted to me of all that is;
> how the world is ordered, what influence have the elements,
> how the months have their beginning, their middle, and their ending,
> how the sun's course alters and the seasons revolve,
> how the years have their cycles, the stars their places.
>
> To every living thing its own breed, to every beast its own moods;
> the winds rage, and men think deep thoughts;
> the plants keep their several kinds, and each root has its own virtue;
> all the mysteries and all the surprises of nature were made known to me;

---

[174] Knox, Proverbs 8:22-33; Jaki, *Savior of Science*, 67.
[175] Jaki, *Science and Creation*, 146.
[176] Jaki, *Science and Creation*, 153, 161.

wisdom herself taught me, that is the designer of them all.[177]

The Hellenistic Jews held a sacred respect for the Two Books of Maccabees where the first biblical appearance of the phrase creation *ex nihilo* is found.[178] It is the story of the mother who was martyred after watching her seven sons be tortured and martyred first. The sons were tortured as she watched because they refused to break God's command and eat the flesh of swine. Their tongues were cut out, scalps torn off, hands and feet mutilated, while the mother and remaining brothers stood by. Then each one was roasted alive, maimed and suffering as they were. The brothers comforted each other as they died bravely, "God sees true," they said, "and will not allow us to go uncomforted."[179] As they died, the mother continued to hearten her sons:

> Into this womb you came, who knows how? Not I quickened, not I the breath of life gave you, nor fashioned the bodies of you one by one! Man's birth, and the origin of all things, he devised who is the whole world's Maker; and shall he not mercifully give the breath of life back to you, that for his law's sake hold your lives so cheap?[180]

Outraged at the defiance of his authority, the king turned to the youngest and only still-living son whom the mother counseled in her native tongue:

> Nine months in the womb I bore thee, three years at the breast fed thee, reared thee to be what thou art; and now,

---

[177] Knox, Wisdom 7:17-21; Jaki, *Science and Creation*, 154.
[178] Knox, 2 Maccabees 7:28; Jaki, *Savior of Science*, 71.
[179] Knox, 2 Maccabees 7:6.
[180] Knox, 2 Maccabees 7:22-23.

my son, this boon grant me. Look round at heaven and earth and all they contain; bethink thee that of all this, and mankind too, *God made out of nothing*. Of this butcher have thou no fear; claim rightful share among thy brethren in yonder inheritance of death; so shall the divine mercy give me back all my sons at once.[181]

Jaki tied this story to the history of science because it demonstrates the radically different view of creation held by the Old Testament cultures. He explains, "No martyrdom with a hope of bodily resurrection was ever inspired by a *Demiourgos* whose 'creative' power consisted in the ability to manipulate the already existing 'formless' matter into actual shapes."[182] The *Demiourgos* (also called Demiurge) is the name for the Maker or Creator of the world in Platonic and Gnostic philosophy. This intensity of martyrdom is relevant to the uniqueness of Jaki's insight into the history of science and the saving Birth of Christ, the core of his theological fullness. The faith in the mercy of God the Creator was not just an intellectual exercise, it was held so strongly that believers would give up their lives before denying the laws and the faithfulness of God.

Nor was the view of the cosmos an independent view held by cultural thinkers trying to reconcile the search for God with what they observed in nature; it was a view that originated from what *God revealed in Scripture* to Moses, to the prophets, and to His people. This view did not suddenly come into Christian thought after the life, death, and resurrection of Christ; it was fulfilled by those events and revelations. Christians already saw the universe as a work of the Creator who is a personal God and not creation itself. They saw the order in the universe originating from the same source as the order of the laws, and the stability of creation

---

[181] Knox, 2 Maccabees 7:27-29.
[182] Jaki, *Savior of Science*, 71.

gave witness to the faithfulness of God who loves all that exists and holds nothing of what He made in abhorrence.[183] He is a God who holds everything in existence and can destroy it in one blow, or interact in the history of mankind in the same manner, because He rules the cosmos and "ordered all things by measure, number, and weight."[184]

---

[183] Knox, Wisdom 11:24; Jaki, *Savior of Science*, 69.
[184] Knox, Wisdom 11:20-21; Jaki, *Science and Creation*, 154.

Early Christianity

Early Christianity

"Let none of you worship the sun. Let no one deify the universe; rather let him seek after the creator of the universe."[185]

The mindset toward nature and the mindset toward religion were united, and this basic, intrinsic psychology was present in the Old Testament as well as in the New, thriving among the early Christians. If an imperfect religious mindset caused the stillbirths for the cyclic eternal "cosmic treadmills" of Egypt, India, China, Babylon, and Greece, then the break from that psychology is most significant. This devotion to the Old Testament worldview is seen, for instance, in the early Church of the Roman Empire's ancient catechesis which contained a striking number of allusions to this same psalm mentioned in the previous section.

> The Lord is my shepherd; how can I lack anything?
> He gives me a resting-place where there is green pasture,
> leads me out to the cool water's brink, refreshed and content.
> As in honour pledged, by sure paths he leads me;
> dark be the valley about my path,
> hurt I fear none while he is with me;
> thy rod, thy crook are my comfort.
> Envious my foes watch, while thou dost spread a banquet for me;
> richly thou dost anoint my head with oil, well filled my cup.
> All my life thy loving favour pursues me;
> through the long years the Lord's house shall be my dwelling-place.[186]

---

[185] Clement of Alexandria, *The Exhortation to the Greeks*, 143; quoted in Jaki, *Science and Creation*, 168.
[186] Knox, Psalm 23.

In the writings of St. Cyril of Jerusalem, St. Ambrose, Didymus of Alexandria, St. Gregory of Nyssa, Origen, and St. Cyprian, among others, a catechetical reference to this psalm is found.[187] The shepherd is of course a reference to Christ. The pasture is the fresh and green words of Scripture that nourishes the hearts of believers and gives them spiritual strength, a place of repose. The cool, still water is the water of Baptism where sin is destroyed and a new creature is born. The sacraments, being protective, lead on a sure path safe from fear or harm from demons. The rod and the staff are understood to be the outpouring of the Holy Spirit who guides. On the Paschal night, the newly-baptized catechumens are led to the table prepared for them, the Eucharist, to assist at Mass for the first time. Their heads are anointed with oil, the sign of the Cross, a mark of protection and of identity. The overflowing cup, the chalice, is the Eucharistic wine, a sober inebriation that fills the heavy and gloomy heart with the joy of divine goodness. All the days of life are a process of conversion, a journey toward the dwelling in the Lord's house forever, a journey in the visible Church, a membership in the people of God, hoping for the Kingdom of Heaven. The same worldview of a rational and pre-scientific psychology was also the worldview of Christian religious initiation. It was a united worldview.

In *Science and Creation*, Jaki devoted the eighth chapter to the writings of early Church Fathers to show that although much of the writing was dedicated to Christian ethics, scriptural exegesis, and theology, there was also a considerable effort to defend the superiority of the Christian message over paganism and pantheism.[188] Jaki noted that the classic textbook *Christianity and Classical Culture: A Study of*

---

[187] Jean Marie Danielou, S. J. *The Bible and the Liturgy* (Notre Dame, IN: University of Notre Dame Press, 2002), 92-98.
[188] Jaki, *Science and Creation*, 163.

Thought and Action from Augustus to Augustine (1944) gave no account of the Church Fathers' attitudes toward Greek and Roman science, and at the time of writing *Science and Creation* (1974), a "careful and comprehensive study of all such texts relevant to the question [was] still wanting" then, and that seems to be the case today as well.[189]

The more modern textbook *Backgrounds of Early Christianity* (1987, 1993, 2003) gave some detail of the differences in the worldview of pantheistic Greek and biblical cultures, including the ways Aristotelian philosophy conflicted with Christian thought, but it still was not the focused study on the scientific attitudes of the Church Fathers that Jaki called for.[190] Given the connection Jaki made between the biblical worldview of Creation out of nothing and the birth of science in the Christian West, a closer evaluation of these texts would be a most beneficial addition to Jaki's thesis that science was born of Christianity.[191] For the purpose of this present work, a brief discussion of the attitudes of some of the Church Fathers is offered and supported with quotes and contexts of the early writings, enough to show that such a study would be warranted.

St. Justin Martyr (c. 100–165 A.D.) made the point in his *First Apology* that Christians should not be hated since their beliefs were similar, but superior to, the Stoic doctrines. In Chapter XX, "Heathen analogies to Christian doctrine," he clarified the distinction: "And philosophers called Stoics teach that even God Himself shall be resolved into fire, and they say that the world is to be formed anew by this

---

[189] Jaki, *Science and Creation*, 188; referring to Charles N. Cochrane, *Christianity and Classical Culture: A Study of Thought and Action from Augustus to Augustine* (Oxford: Oxford University Press, 1944).

[190] Everett Ferguson, *Backgrounds of Early Christianity* (Grand Rapids, MI: William. B. Eerdmans Publishing Co., 1987, 1993, 2003).

[191] This is a possible future research project.

revolution; but we understand that God, the Creator of all things, is superior to the things that are to be changed."[192] In his *Second Apology to the Roman Senate*, he tried to explain why the Stoic morality did not hold under the doctrine of eternal cycles:

> For if they say that human actions come to pass by fate, they will maintain either that God is nothing else than the things which are ever turning, and altering, and dissolving into the same things, and will appear to have had a comprehension only of things that are destructible, and to have looked on God Himself as emerging both in part and in whole in every wickedness; or that neither vice nor virtue is anything; which is contrary to every sound idea, reason, and sense.[193]

This quote is to show that in the first century, just as in biblical times, just as in the next millennium, and just as still today, the Christian dogma of creation out of nothing by the Creator who is distinct and separate from His creation has always been a different worldview with radically different logical conclusions about the cosmos. Either the world is eternal and God and man alike are caught up in the cycles, or the world has a beginning and an end in a universe that is ordered by the Creator.

Another apologist, Athenagoras (ca. 133–190 A.D.), made the distinction between God and creation. He taught that Christians, not the pagans, were the ones "who distinguished God from matter, and teach that matter is one thing and God

---

[192] Alexander Roberts, Sir James Donaldson, Arthur Cleveland Coxe, editors, *Ante-Nicene Fathers Volume I: The Apostolic Fathers, Justin Martyr, Irenaeus* (New York: Charles Scribner's Sons, 1925), 169; quoted in Jaki, *Science and Creation*, 164.

[193] Roberts, *Ante-Nicene Fathers Volume 1*, 191; quoted in Jaki, *Science and Creation*, 165.

another, and that they are separated by a wide interval, for the Deity is uncreated and eternal, to be beheld by the understanding and reason alone, while matter is created and perishable . . . "[194] He also taught that the world was "an instrument in tune, and moving in well-measured time," and that the Deity is the only one who deserved worship because He gave the world "its harmony, and strikes its notes, and sings the accordant strain."[195] Athenagoras, consistent with his fellow Christians of the future Middle Ages, noted that the failure of philosophers to realize this distinction led them into inconsistencies about the origin and permanence of the world. He cited Plato and the Stoics as having such an inconsistency, noting that the eternal and cycling conception of the cosmos contradicts the very idea of a God who is the Creator because there is no account for how the providential and governing cause can exist without the passive and changeable cause.

> Discoursing of the intelligible and the sensible, Plato teaches that that which always is, the intelligible, is unoriginated, but that which is not, the sensible, is originated, beginning to be and ceasing to exist. In like manner, the Stoics also say that all things will be burnt up and will again exist, the world receiving another beginning. But if, although there is, according to them, a twofold cause, one active and governing, namely providence, the other passive and changeable, namely matter, it is nevertheless impossible for the world, even though under the care of Providence, to remain in the

---

[194] Alexander Roberts, Sir James Donaldson, Arthur Cleveland Coxe, editors, *Ante-Nicene Fathers Volume II: Fathers of the Second Century: Hermes, Tatian, Athenagoras, Theophilus, and Clement of Alexandria* (New York: Charles Scribner's Sons, 1925) 131; quoted in Jaki, *Science and Creation*, 164.

[195] Roberts, Vol. II, 136. quoted in Jaki, *Science and Creation*, 164.

same state, because it is created—how can the constitution of these gods remain, who are not self-existent, but have been originated? And in what are the gods superior to matter, since they derive their constitution from water? But not even water, according to them, is the beginning of all things. From simple and homogeneous elements what could be constituted? Moreover, matter requires an artificer, and the artificer requires matter. For how could figures be made without matter or an artificer? Neither, again, is it reasonable that matter should be older than God; for the efficient cause must of necessity exist before the things that are made.[196]

St. Irenaeus (A.D. 2nd century–c. 202) recounted heresies in his monumental work, *Against Heresies*. The ideas of the Valentinians, Marcosians, Nicolaitans, Encratites, Borborians, Ophites, Sethians, and others listed were pagan forms of Christianity, efforts to reconcile both thoughts. They led to such ideas as dualism, demonology, fatalism, reincarnation, emanationism, and pantheism fused with Christian details.[197] Irenaeus' purpose was to teach the faithful, and to do this he emphasized two points: 1) that faith in the Creator of all was the basis of Christian belief, and 2) that only a firm adherence to the Church could guard the intellectual vision derived from that faith.[198] In order to move on in his chapter having noted that deeper studies might ensue, Jaki summarized the earliest Christian apologists as such: "To mention after all this Irenaeus' insistence on a creation out of nothing, or his needling of the inner contradictions of the eternal recurrence

---

[196] Roberts, Vol. II, 136. quoted partially in Jaki, *Science and Creation*, 164.

[197] Jaki, *Science and Creation*, 165.

[198] Jaki, *Science and Creation*, 165.

through reincarnations, would be to belabor already familiar points."[199]

As Christianity spread rapidly throughout the Roman Empire, Christian thought and fundamental characteristics of Greek science achieved a "sophisticated awareness" crystallizing in Alexandria where the first school of Christian thought emerged at the Alexandrian School which had been founded by Alexander the Great in 322 B.C.[200] Clement of Alexandria (died A.D. 215) was an intellectual who studied with Christian teachers elsewhere before coming to Alexandria to teach at the school and refute paganism and pantheism.[201] One of his students was Origen (c. A.D. 182–251), whose general outlook was affected by the Platonism prevalent in Alexandria where he was born. Origen is known his systematic study of Scripture, his theological cosmology, and for his refutation of paganism.[202] Clement and Origen had a "double task" of fully expressing the existing Christian intellectual tradition in that they had to articulate the Covenant to the faithful and serve as apologists to the pagan world, which required them to address the cosmology.[203]

To set the tone of the time of Clement and Origen, there is a quote found at the end of *The Pastor of Hermas* (sometimes called the *Shepherd of Hermas*) which was so valued a book in the first and second century as to be considered canonical scripture by some of the Fathers. There is clearly a strong emphasis on the same biblical worldview of the Old Testament cultures and the same worldview of the Middle Ages when science was born, a worldview that asserted the

---

[199] Jaki, *Science and Creation*, 165.

[200] "Church History Study Helps: The Alexandrian School," Theology Website at http://www.theologywebsite.com/history/alexandria.shtml.

[201] "Church History Study Helps: The Alexandrian School."

[202] "Church History Study Helps: The Alexandrian School."

[203] Jaki, *Science and Creation*, 163.

power and wisdom of God, an ordered cosmos that He created, with laws of nature and laws for man to live by.

> Lo, the God of powers, who by His invisible strong power and great wisdom has created the world, and by His glorious counsel has surrounded His creation with beauty, and by His strong word has fixed the heavens and laid the foundations of the earth upon the waters, and by His own wisdom and providence has created His holy Church, which He has blessed, lo! He removes the heavens and the mountains, the hills and the seas, and all things become plain to His elect, that He may bestow on them the blessing which He has promised them, with much glory and joy, if only they shall keep the commandments of God which they have received in great faith.[204]

In his *Exhortation to the Greeks*, Clement taught that a result of idol worship was the mental chaining of the intellect to the blind forces of nature, and he refused to follow the pagans in regards to motion and the idolizing of natural forces:

> Far in deed are we from allowing grown men to listen to such tales. Even to our own children, when they are crying their heart out, as the saying goes, we are not in the habit of telling fabulous stories to soothe them; for we shrink from fostering in the children the atheism proclaimed by these men, who, though wise in their own conceit, have no more knowledge of the truth than infants. Why, in the name of truth, do you show those who have put their trust in you that they are under the

---

[204] Roberts, *Volume II: Fathers of the Second Century: Hermes, Tatian, Athenagoras, Theophilus, and Clement of Alexandria* (New York: Charles Scribner's Sons, 1925), Book I, Chapter iii.

dominion of "flux" and "motion" and "fortuitous vortices"? Why, pray, do you infect life with idols, imagining winds, air, fire, earth, stocks, stones, iron, this world itself to be gods? Why babble in high-flown language about the divinity of the wandering stars to those men who have become real wanderers through this much-vaunted,—I will not call it astronomy, but— astrology?[205]

Clement urged for a more confident attitude toward nature, a view of a world created by a rational Creator. Not only did he exhort the Greeks to view the world as creation, a robust confidence in human and cosmic existence, but he exhorted them to have faith in Christ who generated that confidence:

How great is the power of God! His mere will is creation; for God alone created, since He alone is truly God. By a bare wish His work is done, and the world's existence follows upon a single act of His will. Here the host of philosophers turns aside, when they admit that man is beautifully made for the contemplation of heaven and yet worship the things which appear in heaven and are apprehended by sight. For although the heavenly bodies are not the works of man, at least they have been created for man. Let none of you worship the sun. Let no one deify the universe; rather let him seek after the creator of the universe.[206]

---

[205] Clement of Alexandria, translated by G. W. Butterworth, *The Exhortation to the Greeks, The Riches of Man's Salvation, and the Fragment of an Address Entitled To the Newly Baptized* (London: William Heinemann, 1919), chapter vi, 153; quoted in Jaki, *Science and Creation*, 168.

[206] Clement of Alexandria, *The Exhortation to the Greeks*, 143; quoted in Jaki, *Science and Creation*, 168.

Origen tried, especially in his *De Principiis* (*On First Principles*), to synthesize Christianity with pagan and Eastern ideas of the cosmos, and he sought understanding of the eternal cycles. He wrote:

> So therefore it seems to me impossible for a world to be restored for the second time, with the same order and with the same amount of births, and deaths, and actions; but that a diversity of worlds may exist with changes of no unimportant kind, so that the state of another world may be for some unmistakable reasons better (than this), and for others worse, and for others again intermediate. But what may be the number or measure of this I confess myself ignorant, although, if anyone can tell it, I would gladly learn.[207]

His searching attitude notwithstanding, Origen still noticed the impossibility of eternally repeating worlds, and he noticed that such an idea was in conflict with revelation. He recalled the events of biblical and salvation history, noting that if the world repeated itself over and over again, then there would be more than one Adam and Eve, more than one Deluge, more than one Moses, more than one Judas and Paul, and more than one Savior. If this were true, he furthered, and everything man did in this age were to be repeated, then there would be no free will because souls driven in a cycle to be endlessly repeated are, thus, all predetermined.

---

[207] Alexander Roberts, Sir James Donaldson, Arthur Cleveland Coxe, editors, *Ante-Nicene Fathers Volume IV: Tertullian, Part Fourth; Minucius Felix; Commodian; Origen, Part First and Second* (New York: Charles Scribner's Sons, 1925), Origen, *De Principiis*, Book II, Chapter 3 "On the Beginning of the World, and Its Causes," paragraph 4, 273.

And now I do not understand by what proofs they can maintain their position, who assert that worlds sometimes come into existence which are not dissimilar to each other, but in all respects equal. For if there is said to be a world similar in all respects (to the present), then it will come to pass that Adam and Eve will do the same things which they did before: there will be a second time the same deluge, and the same Moses will again lead a nation numbering nearly six hundred thousand out of Egypt; Judas will also a second time betray the Lord; Paul will a second time keep the garments of those who stoned Stephen; and everything which has been done in this life will be said to be repeated—a state of things which I think cannot be established by any reasoning, if souls are actuated by freedom of will, and maintain either their advance or retrogression according to the power of their will. For souls are not driven on in a cycle which returns after many ages to the same round, so as either to do or desire this or that; but at whatever point the freedom of their own will aims, there do they direct the course of their actions.[208]

He invoked the authority of Holy Scripture with the expression of the Lord, "I will that where I am, these also may be with Me; and as I and You are one, these also may be one in Us."[209] He pondered that this "may not seem to convey something more than an age and ages," but he also noted that there is an end to time when "all things are now no longer in an age, but when God is in all."[210]

At the conclusion of a long discourse against Celsum, a Platonist, Origen reiterated a firm conviction that the cosmic

---

[208] Roberts, *Ante-Nicene Fathers Volume IV*, 273; quoted in Jaki, *Science and Creation*, 171.

[209] Roberts, *Ante-Nicene Fathers Volume IV*, 273.

[210] Roberts, *Ante-Nicene Fathers Volume IV*, 273.

vision was not predicated on eternal cycles but on the fusion of truth and benevolence, a central factor in the Christian message, the recognition that Jesus Christ is the Incarnate Word of God.[211] There is no place for the resurrection in the doctrine of cosmic cycles, and the early Christian Fathers recognized this clearly.

> For we know that even if heaven and earth and the things in them pass away, yet the words about each doctrine, being like parts in a whole or forms in a species, which were uttered by the *Logos* who was the divine *Logos* with God in the beginning, will in no wise pass away. For we would pay heed to him who says: "Heaven and earth shall pass away, but my words shall not pass away."[212]

Origen, like many of the early Church Fathers, demonstrated the depth of his conviction by martyrdom.[213] Just as Jaki noted in the story of the mother whose seven sons were martyred in the book of Maccabees, the worldview of the Bible and of Christianity was not merely a philosophical outlook; it was a pervasive conviction that was kept pure and protected at any price because the faithful held it *as true*.

Finally, a survey of the early Church and scientific attitudes born of the Christian faith cannot be complete without a mention of St. Augustine of Hippo (A.D. 354–430). In his work *The City of God*, Augustine systematically addressed questions about the destiny of man, which hardly made sense in the doctrine of eternal cycles.[214] According to

---

[211] Jaki, *Science and Creation*, 175.

[212] Origen, translated by Henry Chadwick, *Contra Celsum* (Cambridge: University Press, 1953), 281; quoted in Jaki, *Science and Creation*, 175.

[213] Jaki, *Science and Creation*, 175.

[214] Jaki, *Science and Creation*, 177.

Jaki, Augustine's work "moulded more than any other book by a Christian author the spirit of the Middle Ages" because its "pages were as many wellsprings of information and inspiration for the emerging new world of Europe about the meaning of mankind's journey through time."[215] Fundamentally and consistent with Christian teaching before him, Augustine taught that the physical universe had its origin in the sovereign act of creation by God. It became the "intellectual vehicle for a confidence which centuries later made possible the emergence for the first time of a culture with a built-in force of self-sustaining progress."[216] *The City of God* begins with six chapters that extensively lay out the considerations about Creation, the finiteness of the universe in time and space, and the goodness and beauty of the universe issued by the Creator Himself. It was baffling to Augustine that anyone would believe that good is not the source of all things.

> But it is much more surprising that some even of those who, with ourselves, believe that there is one only source of all things, and that no nature which is not divine can exist unless originated by that Creator, have yet refused to accept with a good and simple faith this so good and simple a reason of the world's creation, that a good God made it good; and that the things created, being different from God, were inferior to Him, and yet were good, being created by none other than He.[217]

---

[215] Jaki, *Science and Creation*, 177.

[216] Jaki, *Science and Creation*, 178.

[217] Philop Schaff, editor, *A Select Library of the Nicene and Post-Nicene Fathers of the Christian Church*, St. Augustine, *The City of God*, Volume II, "St. Augustine's City of God and Christian Doctrine," (Grand Rapids, MI: William. B. Eerdmans Publishing Co., 1956), Book XI, Chapter 23, 217.

It was also incomprehensible to him how anyone could be satisfied with the doctrine of eternal cycles alternating forever between birth and decay, happiness and misery. He preferred instead a sound doctrine and a straight path prescribed by the Creator to all of Creation, His Handiwork. In Book XII, Chapter 13 "Of the Revolution of the Ages, Which Some Philosophers Believe Will Bring All Things Round Again, After a Certain Fixed Cycle, to the Same Order and Form as at First," Augustine was clear that such a view was a "transmigration between delusive blessedness and real misery."[218]

> For how can that be truly called blessed which has no assurance of being so eternally, and is either in ignorance of the truth, and blind to the misery that is approaching, or, knowing it, is in misery and fear? Or if it passes to bliss, and leaves miseries forever, then there happens in time a new thing which time shall not end. Why not, then, the world also? Why may not man, too, be a similar thing? So that, by following the straight path of sound doctrine, we escape, I know not what circuitous paths, discovered by deceiving and deceived sages.

When other scholars tried to interpret biblical references as evidence of eternal cycles, Augustine strongly rejected such an interpretation, just as his predecessors had, on the grounds of the impossibility of more than one Savior:

> At all events, far be it from any true believer to suppose that by these words of Solomon those cycles are meant, in which, according to those philosophers, the same periods and events of time are repeated; as if, for example, the philosopher Plato, having taught in the school at Athens which is called the Academy, so,

---

[218] Augustine, *The City of God*, Book XII, Chapter 13, 234.

numberless ages before, at long but certain intervals, this same Plato and the same school, and the same disciples existed, and so also are to be repeated during the countless cycles that are yet to be—far be it, I say, from us to believe this. For once Christ died for our sins; and, rising from the dead, He dies no more. Death has no more dominion over Him; (Romans 6:9) and we ourselves after the resurrection shall be ever with the Lord (1 Thessalonians 4:16) to whom we now say, as the sacred Psalmist dictates, You shall keep us, O Lord, You shall preserve us from this generation. And that too which follows, is, I think, appropriate enough: The wicked walk in a circle, not because their life is to recur by means of these circles, which these philosophers imagine, but because the path in which their false doctrine now runs is circuitous.

Augustine had an appreciation for quantitative relationships, but that was not his main concern. His main concern was the quest for happiness. His view was that knowledge of the quantitative exactness of the natural world, including the cosmos, could not help much in understanding the biblical message.[219] Augustine also rejected any biblical interpretation which denied or ignored the established conclusions of natural studies. He was explicit on this point:

It is often the case that a non-Christian happens to know something with absolute certainty and through experimental evidence about the earth, sky, and other elements of this world, about the motion, rotation, and even about the size and distances of stars, about certain defects [eclipses] of the sun and moon, about the cycles of years and epochs, about the nature of animals, fruits, stones, and the like. It is, therefore, very deplorable and

---

[219] Jaki, *Science and Creation*, 182.

harmful, and to be avoided at any cost that he should hear a Christian to give, so to speak, a "Christian account" of these topics in such a way that he could hardly hold his laughter on seeing, as the saying goes, the error rise sky-high.[220]

Augustine realized that when statements of the Bible conflicted with hypotheses of the workings of nature and when reason and observation provided no clear solution and decisive evidence, nor did Scripture seem to be explicitly literal, then the matter was open to further inquiry. Whenever scientific reasoning seemed to settle a matter, however, he urged that Scripture would have to be reinterpreted. When an apparent conflict could not be settled, he wrote that questions which "require much subtle and laborious reasoning to perceive which the actual case" he had no time for because "it is not needed by those whom [he wished] to instruct for their own salvation and for the benefit of the Church."[221] Jaki wrote of Augustine's contributions to next millennium:

A man with a restored sense of purpose, a man with an ability to discern intelligible patterns in the universe, a man aware of the vital difference between knowledge and happiness, a man confronting an external world not as an a priori product of his mind but accessible to the light of reason which itself was a participation in God's mind, such were some principal consideration which Augustine

---

[220] J. Zycha, editor, Augustine, *Sancti Aureli Augustini De Genesi ad litteram libri duodecim*, in *Corpus Scriptorum Ecclesiasticorum Latinoram*, Volume XXVIII, Section III, Part 1 (Vienna: F. Tempsky, 1894), Book I, Chapter 19, 28-29; quoted in translation to English in Jaki, *Science and Creation*, 182.

[221] *De Genesi*, Book II, Chapter 10, 47 quoted in translation to English in Jaki, *Science and Creation*, 183.

stressed through his literary career as a thinker and a Christian.[222]

For another thousand years, the writings and wisdom of Augustine would remain a principal source of instruction that held consequences for the coming new phase of human history immersed in scientific enterprise.

---

[222] Jaki, *Science and Creation*, 183.

The Christian West

The Christian West

"I might seek from the theological masters what they might teach me in these matters as to how these things take place…"

This chapter "Was Born" will now turn to the Christian scholars of the Middle Ages. They are discussed in chronological order to show the progression toward the breakthrough of science, a breakthrough dependent on divine revelation, which gave birth to modern science. This review of ten scholars seeks to show this progression away from the pantheistic ideas received from the Greek scientific *corpus* toward the insight that secured the viable breakthrough, or birth, of modern science.

Adelard of Bath

It was under the stronghold of faith in a Creator from Old Testament times and strengthened through the first millennium of Christianity that the European scholars received the Greek philosophical and natural works from the Arabs. The birth of science in Europe begins around the twelfth century with the writing of Adelard of Bath (1080–1125), well into the so-called Dark Ages identified by some historians.[223] The story, if the writings of the philosophers are taken into account, is much more complex than Christian medieval thought needing a mathematical and technological trigger to give birth to modern science.

Jaki denoted Adelard's *Questiones naturales* as the "true dawn of medieval science" because it starts with phrases that are "pregnant with the future."[224] Amid a genuine devotion to the miraculous during the European Middle Ages, Adelard's nephew, who studied in France at his uncle's school, wanted

---

[223] Jaki, *Science and Creation*, 219.
[224] Jaki, *Science and Creation*, 219.

Adelard to publish something fresh from the learning he gained during his travels in the Muslim world.[225] Adelard translated texts of Arab trigonometry, astronomy, and Euclid's geometry into Latin and was, therefore, familiar with the struggles of Arab thinkers to reconcile faith and reason.[226]

Jewish, Muslim, and Christian thinkers had in common a conflict between Greek and Eastern ideas based on eternal cosmic cycles and the belief in a personal Creator who created the universe in an absolute beginning of time. Adelard was aware of this conflict when he answered his nephew's questions. In fact, the nephew's questions demonstrated his awareness of this struggle and the how the Christian faith kept a *realistic* view of the world firmly in place. The younger had a lot of questions for his uncle. He wanted to know why plants spring from the earth.[227] He wanted to know the cause and how it is explained. He wanted to know how trees burst through the ground and put out branches. He wanted to know why plants spring up even if he put dry dust in a bronze pot. The youth was inclined to attribute all of these things to the miraculous effect of the divine will. Nature, after all, has plenty of seeming miracles to feed the mind in search of them. However, Adelard urged his nephew to find a balance between faith and reason, to use reason as far as possible.

He admitted that it was the will of the Creator that plants should spring from the earth, but he also asserted that there was a *reason* (a cause) for it. When his nephew asked if it were

---

[225] Adelard of Bath, *Conversations with His Nephew: On the Same and the Different, Questions on Natural Science and On Birds*, Edited and translated by Charles Burnett (Cambridge, UK: Cambridge University Press, 1998), 83.

[226] Jaki, *Science and Creation*, 219.

[227] Alistair Cameron Crombie, *The History of Science from Augustine to Galileo* (Mineola, NY: Dover Publications, Inc.: 1995 reprint), 45.

not "better to attribute all the operations of the universe to God," Adelard replied:

> I do not detract from God. Whatever this is, is from Him and through Him. But the realm of being is not a confused one, nor is it lacking in disposition which, so far as human knowledge can go, should be consulted. Only when reason totally fails, should the explanation of the matter be referred to God.[228]

The historian Alistair Cameron Crombie called this remark the point when the "medieval conception of nature began to cross the great watershed that divides the period when men looked to nature to provide illustrations for moralizing from that in which men began to study nature for its own sake."[229] Consistent with Jaki's thesis, this statement demonstrates the naturalism of the Christian mindset, a realistic naturalism necessary for the vitality of scientific progress. The historian David C. Lindberg also noted that this naturalism was one of the "salient features of twelfth-century natural philosophy" and was not limited to Adelard, but was also in more general treatises by scholars such as William of Conches, Honorius of Autun, Bernard Sylvester, and Clarembald of Arras, most of whom were also associated with schools in France.[230] There was a growing awareness of natural order and physical laws during the Middle Ages.

Jaki's theological commentary on the psychology is interesting. There was an undeniable tendency toward all things miraculous in the Middle Ages as the scholars tried to

---

[228] Crombie, *Augustine to Galileo*, 45; Jaki, *Science and Creation*, 219.
[229] Crombie, *Augustine to Galileo*, 4.
[230] David C. Lindberg, *The Beginnings of Western Science: The European Scientific Tradition in Philosophical, Religious, and Institutional Context, Prehistory to A.D. 1450* (Chicago, IL: The University of Chicago Press, 1992), Chapter 9, 210-211.

incorporate the ancient Greek writings received and translated from the Muslim world. There was a fascination with the magical and astrological among the pantheistic, agnostic views found in those works. Viewing nature through the lens of Christian faith is not something that can be imposed, even if the state tried to impose it through unevangelical methods, as was done in the high Middle Ages. It is one thing to profess and practice a faith outwardly, but another thing altogether to give free assent of the intellect internally. This is why not all Christian scholars chose to view the cosmos through the faith expressed in the Creed, just as the Muslim scholars chose not to view the cosmos according to the Koran.

A person had essentially two options. To yield to a faith in a transcendental God as Creator was to yield to a belief in man's freedom, but that came with tensions and dynamics that were intellectually difficult to deal with. The freedom of human intellect is hampered by the body, and knowledge is neither immediate nor complete on any subject. On the other hand, to acquiesce to a faith in a pantheistic nature-God only required a person to "seek repose" and be "carried effortlessly on the waves of nature's events."[231] Yet this careless repose in nature will ultimately lead to a fight for survival that begs the question about freedom anyway. What is man's purpose and how do his surroundings work? This is why it was important during the medieval times, with the onrush of astrological treatises translated into Latin from Arabic, that the biblical account of creation played a "purifying role." Those who held a firm belief in a transcendental Creator, a creation out of nothing, an absolute beginning and end of time, and the rationality of the universe and of man were able to *reject* the astrogeology and pantheism of the Greeks and Muslims. There was guidance for the

---

[231] Jaki, *Science and Creation*, 220.

Christian scholars to aid them in choosing between these two options.

Thierry of Chartres

Thierry of Chartres (1155) attempted to give a rational account of creation in his *De Septem Diebus et Sex Operum Distinctionibus*.[232] He interpreted the story of creation to mean that God created space and chaos in the beginning and that all things were composed of elements. This naturalness was a departure from Plato's divinity of the heavens and his *Demiurge* that shaped the material world, described in *Timaeus*.[233] That is, he departed from the dichotomy between the heavens and the earth, a view that held the celestial bodies as divine and the terrestrial bodies as material.[234]

Because of his faith, Thierry had no illusion about the difference between the Creator and creation, and he explained the circular motion of the firmament and the stars as a projectile similar to how a "stone is thrown; its impetus is ultimately due to the hold of the thrower against something solid."[235] Classical physics was to be born of such a view towards naturalness of the heavens and this early *impetus* theory. Thierry emphasized the need for mathematics to not only investigate the universe, but also to recognize God. "There are four kinds of reasons," he explained, "that led man to the recognition of the Creator: the proofs are taken from arithmetic, music, geometry, and astronomy."[236] The connection Thierry saw between these four disciplines is consistent with the Old Testament worldview that the Creator arranged everything according to number, measure,

---

232 Crombie, *Augustine to Galileo*, 46.
233 Jaki, *Science and Creation*, 221.
234 This achievement is usually attributed to Galileo, however.
235 Jaki, *Science and Creation*, 221, cited within.
236 Jaki, *Science and Creation*, 221.

and weight. Man's understanding of the physical world had to have a mathematical character to it, mathematics applied to moving objects.

Robert Grosseteste

These developing concepts, Platonic but biblical, were contained in the work of Robert Grosseteste, Bishop of Lincoln (1168–1253). His methodology was rooted in the notion of God as Creator, to deal with points of conflict with Aristotelian philosophy on behalf of Christianity. His thought was strongly shaped by Platonic ideas, but he rejected the idea that the universe emanated from God, as light emanates from the sun, and was instead in favor of the biblical account of creation *ex nihilo*.[237] He made a distinction between the abstract, geometrical notion of math and math as applied to nature. In *De Lineis*, he explains his analysis of the powers propagated from natural agents:

> The utility of considering lines, angles and figures is huge, because it is impossible to know the philosophy of Nature without them. They are valid for the entire universe and, unconditionally, for all its parts. They apply in connecting the properties, such as in straight and circular motions. And they apply in action and passion (reaction), and this is so, whether in the matter or in the capacities of perception; and this is so again, whether in the sense of sight, as it is occurring, or in any other sense in the action of which, it is necessary to add other things on that which is producing the vision.

> For, it does not act through deliberation and choice; and therefore in one way it acts, whatever it is occurring, whether it is a perception or something else, animated or

---

[237] Lindberg, 234.

inanimate. But because of the diversity of the objects of action we have different effects. Moreover, in the perception, this received power produces, in some way, a spiritual and noble effect; on the other hand, when acting on the matter, it produces a material effect, such as the sun produces, through the same power, different effects in different objects of its action. For it hardens the clay and melts the ice.[238]

Grosseteste's theory of scientific measurement was summarized by William of Alnwick from the Oxford Franciscan House in 1310. He understood Grosseteste to hold that all man-made measurements are imperfect since they are based on conventional units which cannot account for the infinitely small.

For how are we to know the number or quantity of that line which the first Measurer has measured? That quantity he reveals to no man, nor can we measure the line by means of infinite points, because they are neither known nor determined to us, as they are to God by whom they are comprehended. Whence this method of measuring is for us as uncertain as the first . . . Therefore there is no perfect measure of continuous quantity except by means of indivisible continuous quantity, for example by means of a point, and no quantity can be perfectly measured unless it is known how many indivisible points it contains. And since these are infinite, therefore their number cannot be known by a creature but by God alone,

---

[238] Amelia Carolina Sparavigna, "Reflection and refraction in Robert Grosseteste's *De Lineis, Angulis et Figuris*" (Cornell University) at http://arxiv.org/ftp/arxiv/papers/1302/1302.1885.pdf; Jaki, *Science and Creation*, 222; Crombie, *Medieval and Early Modern Science*, 20.

who disposes everything in number, weight and measure.[239]

The last line shows that the role of the Creator was in the forefront of Grosseteste's thinking, and there is further evidence of this in his (still unedited) treatises, the "*Hexaemeron*" and the "*De universitatis machina*."[240] Grosseteste struggled with Aristotle's *Physics*, particularly the thoughts on light, and denounced forcibly the Aristotelian idea that the world had no beginning in time.[241] The central theme of *Hexaemeron* was the biblical theme that God is light, truly, essentially and not in the metaphorical sense.[242] Both treatises document that Grosseteste's scientific methodology depended on the Old Testament understanding of the Creator as a rational and personal planner, builder, and maintainer of the universe.[243]

William of Auvergne

There was not a uniform agreement in the thirteenth century about the nature of the universe, however. There were a mixture of opinions and insights from those who studied the Greek works, a mixture of sound principles and irrational tales, reason, and magic.[244] There is evidence of these struggles in the massive works of the Bishop of Paris, William of Auvergne (1180–1249). William took an active interest in the institution where he taught, the University of Paris, and

---

[239] Alistair Cameron Crombie, *Robert Grosseteste and the Origins of Experimental Science 1100-1700* (Oxford: Clarendon Press, 1953), 102-103; Jaki, *Science and Creation*, 223.

[240] Jaki, *Science and Creation*, 223.

[241] James McEvoy, *Robert Grossteste* (Oxford: Oxford University Press, 2000), 85.

[242] McEvoy, 91.

[243] Jaki, *Science and Creation*, 223.

[244] Jaki, *Science and Creation*, 223.

his treatises were aimed at adopting and adapting Aristotle's philosophy on the basis of Christian dogma.[245] His work, therefore, is known as an unexpectedly detailed picture of the magic and superstition of the time.[246] There were errors and perversions of the Aristotelian texts from the Arab translations, so Auvergne's task was to rescue Aristotelian thought from the Arabs by reconciling it with Augustinian and Platonic elements.[247]

There are many references to the magical elements of Auvergne's work, as he sought to find a way out of the mixture of ideas which included the doctrine of the Great Year, the heresy of Manichaeism (the Persian belief that all religious systems could be synthesized, particularly Zoroastrian Dualism, Babylonian folklore, Buddhist ethics, and some superficial tenets of Christianity), fatalism, pantheism, star worship, necromancy, and other irrational concepts.[248] He discussed at length the demonic aspect of magic, superstition, and idolatry.[249]

Lynn Thorndike explained in his book *A History of Magic and Experimental Science* that Auvergne's account of magic is a "remarkable and illuminating one."[250] Most of Auvergne's discourse is found in the last part of the *De universe*. Thorndike summarized the main points:

---

[245] William Turner, "William of Auvergne," *The Catholic Encyclopedia* (New York: Robert Appleton Company, 1912) at New Advent, http://www.newadvent.org/cathen/15631c.htm.

[246] Lynn Thorndike, *A History of Magic and Experimental Science*, Volume 11 of 14 (New York: Columbia University Press, 1923), 338.

[247] Turner, "William of Auvergne."

[248] John Arendzen, "Manichæism," The Catholic Encyclopedia (New York: Robert Appleton Company, 1910) at New Advent, http://www.newadvent.org/cathen/09591a.htm; Jaki, *Science and Creation*, 223.

[249] Thorndike, *A History of Magic*, 344.

[250] Thorndike, *A History of Magic*, 342.

He [Auvergne] constantly assumes that its [magic] great aim is to work marvels. He holds that often the ends are sought by the help of the demons and methods which are idolatrous. Evil ends are brought about by magicians. On the other hand the apparent marvels are often worked by mere human sleight-of-hand or other tricks and deceptions of the magicians themselves. But the marvel can be neither human deceit nor the work of an evil spirit. It may be produces by the wonderful occult virtues resident in certain objects of nature. To marvels wrought in this manner William [of Auvergne] applies the name "natural magic," and has no doubt of its truth. But he denies the validity of many methods and devices in which magicians trust, and contends that marvels cannot be so worked unless demons are responsible.[251]

In the quagmire of ideas there was, nonetheless, an equally pervasive presence of Christian doctrine regarding the Creator and creation. This widespread presence served as a firm barrier against the conceptual chaos that threatened medieval thought.[252] Thorndike also noted this respect for science as an explanation for both natural causes and the Creator's omnipotence:

William at any rate has respect for natural philosophy and favors scientific investigation of nature. Like his namesake of Conches in the preceding century he has no sympathy with those who, when they are ignorant of the causes of natural phenomena and have no idea how to investigate them, have recourse to the Creator's omnipotent virtue and call everything of this sort a

---

[251] Thorndike, *A History of Magic*, 342-343. Thorndike references the original texts of William of Auvergne.
[252] Jaki, *Science and Creation*, 223.

miracle, or evade the necessity of any natural explanation by affirming that God's will is the sole cause of it. This seems to William an intolerable error, in the first place because they have thus only one answer for all questions, and secondly because they are satisfied with the most remote cause instead of the most immediate one. There is no excuse for thus neglecting so many varied and noble sciences.[253]

Jaki thought that William of Auvergne had the right approach for his times. It would have been a mistake summarily to dismiss the translations from the Arabs or to dismiss Aristotelian philosophy where it strayed from Christian doctrine. William, and others, needed to deal with it in detail and to show respect for Plato and Aristotle since Plato and Aristotle had so great a reputation among scholars.

William knew that the doctrine of the Great Year was also a far reaching worldview incompatible with Christian faith, and he wrote a refutation against the six arguments that had been advanced in support of the Great Year. Those arguments could be accused of repetitious detail perhaps, but in his age, it was probably necessary. Still, the refutation came down to one rebuttal, that even among the observed cycles of nature, the recurrences *were never identical.*

The philosopher Alfred North Whitehead wrote in his book *Science and the Modern World,* a series of eight Lowell Lectures delivered at Harvard University in 1925 succinctly describing the overall rebuttal that emerged among Western European minds:

Obviously, the main recurrences of life are too insistent to escape the notice of the least rational of humans; and even before the dawn of rationality, they have impressed themselves upon the instincts of animals. It is

---

[253] Thorndike, *A History of Magic,* 340.

unnecessary to labour the point, that in broad outline certain general states of nature recur, and that our very natures have adapted themselves to such repetitions.

But there is a complementary fact which is equally true and equally obvious:—nothing ever really recurs in exact detail. No two days are identical, no two winters. What has gone, has gone forever. Accordingly the practical philosophy of mankind has been to expect the broad recurrences, and to accept the details as emanating from the inscrutable womb of things beyond the ken of rationality. Men expected the sun to rise, but the wind bloweth where it listeth.[254]

The error could also be described as a failure to distinguish between primary and secondary causes, to distinguish between the Creator and creation. Here, Jaki noted that obviously "what was needed was an unremitting search coupled with an unconditional respect for some guidelines marking the major pitfalls in reasoning about nature, a contingent entity."[255] This unremitting search and unconditional respect would be the work of St. Thomas Aquinas, student of St. Albertus Magnus.

St. Albertus Magnus

Aquinas' master, Albertus Magnus (St. Albert the Great, 1193–1280), was an enthusiastic proponent of the investigative approach to nature. In Western Christendom, Albertus Magnus was the first to interpret comprehensively Aristotle's philosophy.[256] He was heavily influenced by the

---

[254] Alfred North Whitehead, *Science and the Modern World* (Cambridge: The MacMillan Company, 1925), 5.

[255] Jaki, *Science and Creation*, 225.

[256] Lindberg, 237.

Platonism of the Church Fathers, but he perceived the significance of Aristotle's work. He realized that the role of omnipotent determinators to move celestial bodies would not be easily dismissed, and he treated the question of fate in a brief treatise entitled *De fato.*[257] The work begins with a summary of twenty-one writings from Boethius, Aristotle, Macrobius, Ptolemy, Gregory, and Augustine.[258] In interpreting the views of the Stoics and Muslims through the lens of his Christian faith, Albertus rejected astrology and magic, and instead argued for reason and investigation to go as far as possible in the natural world. For instance, he wrote that "changes of the general state of the elements and the world" are affected by the planets and their spheres, and he rejected the opinion of "certain Arabs" that the imaginations of the intelligences that move the moon cause floods.[259]

Albertus also wrote a complete encyclopedia of philosophical disciplines based on the Aristotelian texts for his students of the Dominican Order. He wanted not only to provide theological studies, but also to prepare his students for philosophical studies such as he had not received. He began the extensive part on natural sciences as follows:

> Our purpose in natural science is to satisfy as far as we can those brethren of our order who for many years now have begged us to compose for them a book on physics in which they might have a compete exposition of natural

---

[257] James A. Weisheipl, O.P., *Albertus Magnus and the Sciences: Commemorative Essays* 1980 (Toronto, CA: Pontifical Institute of Medieval Sciences, 1980), "Human Embryology and Development in the Works of Albertus Magnus" by Luke Demaitre and Anthony A. Travill, 414.

[258] Albert the Great (Albertus Magnus), *De fato* (About Fate), Corpus Thomisticum at http://www.corpusthomisticum.org/xpz.html.

[259] Weisheipl, "The Physical Astronomy and Astrology of Albertus Magnus" by Betsy Barker Price, 179.

science and from which also they might be able to understand correctly the books of Aristotle. Although we do not think we are competent of ourselves to carry out this project, nevertheless, because we do not want to refuse our brethren's request, we have finally accepted this task which we so many times rejected. Overcome by the request of certain of these brethren we have undertaken this work first to the praise of Almighty God, who is the fountain of wisdom and the creator, ordered and governor of nature, and then for the benefit of our brethren, and, finally, for the benefit of all those desirous of learning natural science who may read it.[260]

Not only was Albertus assimilating and modifying the Aristotelian texts, he was synthesizing them into Christian thought and teaching it to his pupils.

St. Thomas Aquinas

St. Thomas Aquinas has sometimes been accused of lacking an appreciation for the experimental method in favor of philosophical reasoning, but this is more a personality trait of his than a commentary on his contemporaries. Aquinas was both a philosopher and a theologian, which was key to his stable synthesis of reason and faith. In fact, his works form the classical balance.[261] In keeping with the approach of Auvergne and Albertus, Aquinas reasoned as far as possible with a generous acceptance of the Aristotelian system to show respect for the scholarship of the time.

Aquinas' first major synthesis was the *Summa Contra Gentiles* (1257), which brought the authority of reason to bear

---

[260] Weisheipl, "St. Albert and the Nature of Natural Science" by Benedict Ashley, 78; Quote is from *Physical*, translation 1, c. 1 (1890-1899) edited and translated by Auguste Borgnet.
[261] Jaki, *Science and Creation*, 225.

on Muslim philosophy. The lengthy beginning dealt specifically with the questions about the Creator, the search for truth, human intellect, and refuted specifically theological aspects tied to Muhammad. Aquinas departed from Aristotelic orthodoxy only where no compromise could be allowed by the Christian Creed.[262]

An example of Aquinas' acceptance of the search for the laws of nature can be found in the brief work *De Motu Cordis* (On the Motion of the Heart), in which he considered what moves the heart and exactly what kind of movement it has.[263] He was not referring to the will of man, but to the actual organ called the heart. The movement of this organ is a push-pull motion, somewhat like a circle, which led Aristotle to suppose that this movement had a supernatural mover since it moved as the celestial bodies do, in circles. Aquinas followed Aristotle's physics insofar as he sought a natural explanation for the movement of the heart, but he deviated from Aristotle in concluding that the movement was sustained by intelligences in the way that celestial bodies are moved in circular motion by intelligences in the divine ether.

Instead Aquinas argued that "the motion of the heart is a natural result of the soul, the form of the living body and principally of the heart."[264] Aquinas from there argued that the heart could not have its own soul moved by desire, reasoning thus. First, he noted how inanimate objects and animate beings both move. The heart moves because it is part of an animal's body, which is why the heart beats faster in response to various sensations and emotions, or in the case of man, even in response to reasoning. Second, he noted how the heart beat begins and ends with the life of the animal, a

---

262 Jaki, *Science and Creation*, 225.

263 Thomas Aquinas, *De Motu Cordis*, Translated by Gregory Froelich, The Dominican House of Studies at http://dhspriory.org/thomas/DeMotuCordis.htm.

264 Aquinas, *De Motu Cordis*.

"continuous movement as long as the animal lives."[265] This reasoning demonstrates how Aquinas, just as his contemporaries, desired to search for natural explanations as far as possible even when the reasoning involved spiritual aspects of being.

Aquinas carried the same approach in the *Summa Theologiæ* (1273), accepting as much of Aristotle's philosophy as possible. For instance, in the 91st Question of the Third Part (Supplement), Aquinas discussed "the quality of the world after the judgment." This discussion is a direct confrontation with Aristotle's cosmology of endless cycles emanating from God.[266] Aquinas accepted that there are natural cycles in the world, but in the end when the world is renewed, corporeal things and man will be renewed too. He wrote that the "philosophers who assert that the movement of the heaven will last forever" are not "in keeping with our faith, which holds that the elect are in a certain number preordained by God, so that the begetting of men will not last forever, and for the same reason, neither will other things that are directed to the begetting of men, such as the movement of the heavens and the variations of the elements."[267]

This reasoning guided by faith is the same sort of reasoning that led Aquinas to assert that the movement of the heavens will cease in the world after the judgment not by a natural cause, but as a result of the will of God. This departure from Aristotelian, Greek, and Muslim cosmology–indeed a departure from any other cosmology at all–shows the ultimate reason for the cosmos to exist in the worldview of Christianity. The cosmos exists by the will of the Creator, and it is subordinate to man's eternal, unique, and supernatural destiny.[268]

---

[265] Aquinas, *De Motu Cordis*.
[266] Jaki, *Science and Creation*, 225.
[267] *ST*, III (Supplement), q. 91, a. 2, *Respondeo*.
[268] Jaki, *Science and Creation*, 226.

Roger Bacon

It is often assumed incorrectly that the Franciscan friar Roger Bacon (1214–1294) stood alone in the Middle Ages in his advocacy of experimental and natural sciences, but his works make more sense when considered in the context of his contemporaries. Bacon lived during a time when scholars were trying to make the most of the Greek scientific *corpus* being recovered and translated rapidly.[269] In addition to his strong support for the experimental method, he also lent support to some of the interpretations of the Arabs regarding astrology, enough to get him into some trouble with Church authorities by 1277 when his work *Opus majus* was condemned as containing "some suspected novelties," and he was imprisoned.[270] Bacon is sometimes credited along with Galileo as the beginner of modern science, and indeed he can be considered a precursor for his appreciation for the experimental method.

> There are two ways of acquiring knowledge, one through reason, the other by experiment. Argument reaches a conclusion and compels us to admit it, but it neither makes us certain nor so annihilates doubt that the mind rests calm in the intuition of truth, unless it finds this certitude by way of experience. Thus many have arguments toward attainable facts, but because they have not experienced them, they overlook them and neither avoid a harmful nor follow a beneficial course. Even if a man that has never seen fire, proves by good reasoning that fire burns, and devours and destroys things,

---

[269] Thorndike, *A History of Magic*, 685.

[270] Lynn Thorndike, "The True Roger Bacon," in *The American Historical Review*, Volume XXI (London: The Macmillan Company, 1916), 245.

nevertheless the mind of one hearing his arguments would never be convinced, nor would he avoid fire until he puts his hand or some combustible thing into it in order to prove by experiment what the argument taught. But after the fact of combustion is experienced, the mind is satisfied and lies calm in the certainty of truth. Hence argument is not enough, but experience is.[271]

Bacon can also be praised for his emphasis on the basic unity, interconnectedness, and interdependence of all branches of learning, a concept critical to the success of the early universities.[272] His views did not conflict with the Church so far as he held, along with the Church Fathers, that the Creator was distinct from creation, and therefore, all knowledge formed ultimately a single body of truth.[273] Even as he considered the Greek scientific *corpus*, he also rightly warned that final and secondary causes were distinctions not to be confused.[274] He argued that man can only have partial knowledge of the universe, and he has been considered a visionary about machines that could speed across land, fly in the skies, or make far away objects appear close.[275] Even though the Chinese had already done so, he is credited with the invention of gun powder, the only real experimental

---

[271] Oliver J. Thatcher, ed., *The Library of Original Sources* (Milwaukee: University Research Extension Co.: 1901), Vol. V: The Early Medieval World, 369-376. Found at "Medieval Sourcebook: Roger Bacon: On Experimental Science 1268," by Fordham University at http://www.fordham.edu/halsall/source/bacon2.asp, scanned by Jerome S. Arkenberg, Department of History, California State Fullerton.

[272] Jaki, *Science and Creation*, 226.

[273] Jaki, *Science and Creation*, 226.

[274] Jaki, *Science and Creation*, 226.

[275] Jaki, *Science and Creation*, 226; Thorndike, *A History of Magic*, 654.

success of the "one whom some called the Father of experimental science."[276]

Bacon was, as already said, not the first medieval man to advocate experimentation, and he is not a lone herald of this aspect of modern science. Rather, he reveals the merits and defects of the movement of his time.[277] It was not possible to consider experimental science separate from magic in those days, and the exploration of this history is the main topic of the life's work of the American historian of medieval science and alchemy, Lynn Thorndike, who was also an expert on Roger Bacon's life and works.[278] Some of the experiments Bacon described were fantastic, such as his elixirs that were said to prolong life, and even though the affects were natural, they seemed magical and, therefore made Christians skeptical.[279] Bacon urged that experimentation is necessary:

> First one should be credulous until experience follows second and reason comes third . . . At first one should believe those who have made experiments or who have faithful testimony from others who have done so, nor should one reject the truth because he is ignorant of it and because he has no argument for it.[280]

Bacon's writing about "magic" should be understood in such a context; experimentation did not always have an explanation, even though his discussion of experimental science amounted to little more than an acknowledgement that experience should serve as a criterion of truth.[281] Before

---

[276] Jaki, *Science and Creation*, 227.

[277] Thorndike, *A History of Magic*, 650.

[278] Thorndike, *A History of Magic*, p. 338; Thorndike, "The True Roger Bacon," 245.

[279] Thorndike, *A History of Magic*, 656-657.

[280] Thorndike, *A History of Magic*, 657, from *Opus Majus*.

[281] Thorndike, *A History of Magic*, 658.

there was knowledge of the limits of chemical compounds, people conducting experiments were seeking certain properties, even if they did not realize that those properties did not exist.

There was also a superstitious component in Bacon's writing. He saw a real connection between experiment and magic, most often cited in his views regarding alchemy and astrology. He believed occult science could make gold better than nature and that astrology could predict the future, as did the Arabs and Greeks, although he also regarded magic as demonic.[282] Still in other passages he suggests that magic is not worthless, and he granted that the books of magicians "may contain some truth."[283] According to Thorndike, Bacon goes about as far as Albertus Magnus in "credulous acceptance of superstition, but [he] will not admit, as Albert does, that such things are magic or very closely related to it."[284]

Bacon also held that "celestial bodies are the causes of generation and corruption of all inferior things," consistent with Aristotle's view of the heavens as regulated by angelic intelligences. This view led Bacon to also accept the astrological arts and medicine including the influence of the stars on a person's health and conduct. He believed that the stars inclined men to bad acts or to good conduct, and while he also held that the individual could resist the influence of the stars, the power of the constellations prevailed in the masses of men. This is essentially the view the Greeks held of the Great Year and the Muslims held of astrology (horoscopy), that cycles in history can be predicted by the heavens.[285] Not surprisingly, the *Opus majus*, Bacon's most important work, contains a section on the application of

---

[282] Thorndike, *A History of Magic*, 658-662.

[283] Thorndike, *A History of Magic*, 662.

[284] Thorndike, *A History of Magic*, 664.

[285] Thorndike, *A History of Magic*, 674-675.

astrology to Church government, which was not well-received by the Church.[286]

Sifting through Bacon's assertions may create the impression that they were a continuation of the "Christian attitude of patristic literature to a certain extent."[287] As Thorndike noted, Bacon was a clergyman writing for other clergymen to "promote the welfare of the Church and of Christianity."[288] Hailed as a Father of science, there is no denying that his mindset was formed by Christian faith. His views were true to the age, but also remarkable in that they demonstrate how Christian faith was purifying the Greek scientific *corpus*, transitioning it toward a viable birth of science and out of the errors that kept it stillborn in previous cultures.

Siger of Brabant

The situation could appear to worsen with the thought of Siger of Brabant (1240–1280's), a "radical Aristotelian" who was a contemporary of Magnus and Aquinas. While Magnus and Aquinas sought to harmonize reason and faith (i.e. philosophy and theology), there were others who heeded not the authority of the Church and began to teach philosophical doctrines in contradiction to Christian theology, as the Muslim followers of Averroes did in their approach to the independent autonomies of philosophy and theology.[289]

Siger of Brabant was the best known leader of such a radical faction. He sought to establish the autonomy of philosophy from theology. In doing so, he taught, loyal to Aristotle, that the world emanates from the First Cause and that the philosopher, speaking as a philosopher, had no

---

[286] Jaki, *Science and Creation*, 227-228.
[287] Thorndike, "The True Roger Bacon," 480.
[288] Thorndike, "The True Roger Bacon," 480.
[289] Lindberg, 244.

alternative but to accept the eternity of the world because that is what, in his view, reason dictates. On the other hand, he made it clear that he personally accepted the doctrine of the creation of the world that his faith demanded.[290] In his work *De aeternitate mundi* (On the Eternity of the World), he claimed:

> Since the Prime Mover and Agent is always actual, not something potential before being actual, it follows that it would always move and act, and produces any given [things] without a mediating motion. Furthermore, on the basis of the fact that [the Prime Mover] is always moving and acting, it follows that no species of being (*species entis*) proceeds to act but that it previously had preceded [it], such that with respect to the same species the [things] that existed return circularly—and laws, opinions, religions, and other [things], so that the lesser [things] cycle in virtue of the cycling of the higher [things], although the memory of the cycling of these [things] does not persist, on account of their antiquity.
>
> However, we say these things while reciting the view of the Philosopher, not asserting them as though they were true.[291]

Whether he was trying to placate the Church authorities or sincerely wanted to separate the truth of reason from the truth of faith is a matter of dispute, but the dangerous implications of this divisive approach came to the attention of Aquinas and others who gave a public denunciation of

---

[290] Lindberg, 244.

[291] Siger of Brabant: *The Eternity of the World*, translated by Peter King at http://individual.utoronto.ca/pking/translations/SIGER.Eternity _of_World.pdf.

such views, specifically in Aquinas' short treatise, *De aeternitate mundi contra murmurantes* (*On the Eternity of the World: Against the Mutterers*). Not a polemicist by nature, Aquinas made it clear that the world is not a self-subsisting being, but that its existence is owed to a Creator, and there had to be a beginning and end to the universe in time.[292] As Jaki put it, "about the message of the Christian dogma on creation there could be no misgiving."

> For the believer this could have presented an almost insuperable temptation to espouse certain considerations, suggesting the finiteness of the world in time, as being conclusive. But the most learned among the faithful took consistently the position that no reason but revelation alone could settle a matter that truly needed settling.[293]

Étienne Tempier

St. Thomas Aquinas died in 1274, amid the hesitation and confusion about the Aristotelian texts from the Muslim world, but his defense of the interdependence of faith and reason was, of course, not in vain because later it became the accepted and established synthesis of Catholic and Aristotelian thought. Aquinas' defense was also given during a time leading up to a dramatic event three years later. In 1277, Étienne Tempier, the Bishop of Paris, issued a list of 219 condemned propositions relating to the Aristotelian texts that were irreconcilable to the Catholic worldview. These propositions were not binding on Catholics, but served as a guide for the scholars at the University of Paris. The firm

---

[292] Michael Heller, *Ultimate Explanations of the Universe* (Berlin: Springer-Verlag, 2009), 136.
[293] Jaki, *Science and Creation*, 229.

judgment largely dealt with the eternity of the world and creation.

The propositions are often referenced by historians of science and summarized, as Jaki did in *Science and Creation*. However, it is instructive to review them as they are (in translation) so that they may be considered on their own merit. Below is a list of some of the propositions, specifically the ones that Jaki mentioned in his summary, condemned at the University of Paris by Tempier:[294]

Proposition 27 asserted that God can make as many worlds as He wills. "That the first cause cannot make more than one world."

Proposition 31 rejected that the heavens are divine. "That there are three principles for celestial things: the subject of eternal motion, the soul of the celestial body, and the first mover, [which moves things] insofar as [it is] desired. This is an error with respect to the first two."

Proposition 32 rejected that the world is an organism. "That the eternal principles are two, namely, the body of the heaven and its soul."

Proposition 66 said that rectilinear motion is possible for planets and stars. "That God could not move the heaven in a straight line, the reason being that He would then leave a vacuum."

Proposition 73 essentially condemned pantheism. "That the heavenly bodies are moved by an intrinsic principle which

---

[294] Arthur Hyman, James J. Walsh, and Thomas Williams, editors, *Philosophy in the Middle Ages: The Christian, Islamic, and Jewish Tradition*, Third Edition (Indianapolis, IN: Hackett Publishing Company, Inc., 2010), 541-549.

is the soul, and that they are moved by a soul and an appetitive power, like an animal. For just as an animal is moved by desiring, so also is the heaven."

Proposition 75 also condemned the animistic view of the universe as if the heavens are made of organs. "That the celestial soul is an intelligence, and the celestial spheres are not instruments of intelligences but rather [their] organs, as the ear and the eye are the organs of a sensitive power."

Proposition 83 safeguarded the doctrine of creation. "That the world, although it was made from nothing, was not newly-made, and although it passed from nonbeing to being, the nonbeing did not precede in duration but only in nature."

Proposition 84 condemned the error that the world is eternal. "That the world is eternal because that which has a nature by which it is able to exist for the whole future has a nature by which it was able to exist in the whole past."

Proposition 85 also guarded against an eternal worldview. "That the world is eternal as regards all the species contained in it, and that time, motion, matter, agent, and receiver are eternal, because the world comes from the infinite power of God and it is impossible that there be something new in the effect without there being something new in the cause."

Proposition 86 protected the reality of time and eternity. "That eternity and time have no existence in reality but only in the mind."

Proposition 87 guarded the absolute beginning and end of time. "That nothing is eternal from the standpoint of its end that is not eternal from the standpoint of its beginning."

Proposition 88 rejected that time is infinite. "That time is infinite at both ends. For even though it is impossible for an infinitude [of things] to have been traversed, some one of which had to be traversed, nevertheless it is not impossible for an infinitude [of things] to have been traversed, none of which had to be traversed."

Proposition 89 declared that it is possible to refute arguments of Aristotle if his arguments for an eternal universe contradict revelation. "That it is impossible to refute the arguments of the Philosopher [Aristotle] concerning the eternity of the world unless we say that the will of the first being embraces incompatibles."

Proposition 90, carefully noted a distinction between Creator and creation. "That the universe cannot stop, because the first agent has [the ability] to transmute in succession eternally, now into this form, now into that one. And likewise, matter is naturally apt to be transmuted."

Proposition 91 asserted the beginning of the universe. "That there has already been an infinite number of revolutions of the heaven, which it is impossible for the created intellect but not for the first cause to comprehend."

Proposition 92 condemned the doctrine of the Great Year. "That with all the heavenly bodies coming back to the same point after a period of thirty-six thousand years, the same effects as now exist will reappear."

Proposition 105 rejected determinism based on the stars, particularly that the arrangement of the stars affects an individual from the moment of birth. "That when a man is generated as to his body, and consequently as to [his] soul, which follows the body, there is a disposition in the man [coming] from the order of superior and inferior causes, [and] inclining [him] to such and such actions or events. [This is] an error unless it is meant [to apply only] in the case of natural events and [for an inclination] by way of disposition."

144

Proposition 107 rejected that celestial bodies are the primary source of all matter. "That God was unable to have made prime matter except by means of a celestial body."

Historians are still studying the writings of contemporaries of this time to ascertain the extent of the drama involved, but these condemnations speak for themselves. They represent a distinct struggle between philosophy and theology, reflective of the tension of the time. For the Church Fathers as well as the medieval scholastics, this view demonstrates that philosophy was a "handmaiden" of theology because faith was above reason, but not unreasonable.

As the Aristotelian texts, unchallenged by Greek or Muslim scholars, were accepted into Christendom, it could only have followed that theologians and philosophers of that time would seek to reconcile the contradictions. It is worth noting that even St. Thomas expressed that the rejection of the eternity of the world was a **matter of faith in divine revelation** and *not* a matter of (what would later be called) scientific demonstration or reason:

> The articles of faith cannot be proved demonstratively, because faith is of things "that appear not" (Hebrews 11:1). But that God is the Creator of the world: hence that the world began, is an article of faith; for we say, "I believe in one God," etc. And again, Gregory says (Hom. i in Ezech.), that Moses prophesied of the past, saying, "In the beginning God created heaven and earth": in which words the newness of the world is stated. Therefore the newness of the world is known only by revelation; and therefore it cannot be proved demonstratively.

By faith alone do we hold, and by no demonstration can it be proved, that the world did not always exist, as was said above of the mystery of the Trinity (32, 1).[295]

Historians have and will continue to look upon this period of time in various ways. The condemnations of the University of Paris could be interpreted, as some historians have done, as antagonistic to the autonomy of philosophy, a symbol of an "intellectual crisis" in the University and culture of the late thirteenth century, and that interpretation is partially correct.[296] There was an intellectual struggle between the proposed truths of Aristotelian philosophy and the divine truths of revelation. While Muslim scholars encouraged this separate autonomy, Christian scholars could not, and that refusal to separate the truths of faith from the truths of reason is of utmost significance in understanding what led to the birth of science in Christian Europe because it is a distinction that isolates that religious culture from all others.

This significance is lost on historians who do not consider the theological history of science. Among the leading authorities of the history of science in the Middle Ages, Edward Grant (b. 1926) acknowledged that there was an "atmosphere of fear and hostility toward Greek science and philosophy" when Christianity "manifested its earliest concern about the physical world."[297] He wrote that the Church Fathers and Christian authors of late antiquity

---

[295] *ST*, I, q. 46, a. 2, *Sed contra.*

[296] Jan A. Aertsen, Kent Emery, Andreas Speer, and Walter de Gruyter, *Nach der Verurteilung von 1277 / After the Condemnation of 1277: Philosophie und Theologie an der Universität von Paris im letzten Viertel des 13. Jahrhunderts. Studien und Texte [Philosophy and Theology at the University of Paris in the Last Quarter of the Thirteenth Century] Studies and Texts* (Berlin, New York: Walter de Gruyter GmbH & Co., 2000), 3.

[297] Edward Grant, *The Nature of Natural Philosophy in the Late Middle Ages* (The Catholic University of America, 2010), 49-50.

"grudgingly came to tolerate [pagan enterprises] as handmaidens to theology."[298] The charges of "intellectual crisis" and "grudging" toleration of pagan opinions seem to suggest that Christian scholars were acting imprudently, but they were not. They were approaching scholarship the same way as their predecessors did. For instance, the Greek apologists of early Christianity, such as Clement of Alexandria and Gregory of Nyssa, welcomed contributions from Greek thought.

However, Pierre Duhem (1861–1916), a French physicist, had a different opinion, an opinion which was probably the catalyst for the ongoing deeper studies into the Condemnations of 1277. In the early twentieth century, Duhem conducted pioneering research on the origins of classical physics, uncovering medieval texts by Christian scholars that were before not included in the historical accounts of the Middle Ages. He is credited by Jaki as "single-handedly" inspiring a "vigorous interest in medieval science."[299] Duhem's research countered and corrected the long-held belief that the Church stifled scientific progress, particularly in the Middle Ages (pejoratively called the "Dark Ages"). The most important historical corrections provided by Duhem's original sources were the anticipations by Christian theologians of the concepts of *inertia* and *momentum*, the search for both qualitative and quantitative analysis of physical processes, and the realization of the importance of experimental investigation to understand nature.[300]

Duhem's discovery of original writings from the Middle Ages successfully demonstrated that the period between 1200 and 1500 in Western Europe was not a dark period of intellectual stagnation, as was erroneously but universally

---

298 Grant, *The Nature of Natural Philosophy*, 49.
299 Jaki, *Science and Creation*, 231.
300 Jaki, *Science and Creation*, 231.

assumed even into the twentieth century.[301] According to Grant, Duhem successfully struck a blow to that myth and demonstrated that this period played a role in the history of science. Duhem went further; he held that these achievements of Middle Age Christendom were essential to producing the Scientific Revolution of the seventeenth century. For Duhem the Condemnations of 1277 marked the birth of modern science.[302] Duhem's birth date for modern science has, of course, been met with rejection among some historians of the Middle Ages and of science. Even Jaki, who called himself not only an admirer but a "kindred spirit," somewhat revised this particular conclusion of Duhem's. Jaki did not mark the birth of modern science as a date in time but as a conceptual breakthrough that freed scholarship from Aristotelian errors and allowed the understanding of physical laws and systems of laws to grow into a viable, self-sustaining discipline fundamentally based on the quantitative measurement of objects in motion.

Jaki credited the Condemnations of 1277 for shaping the mindset and conceptual framework for a revolutionary new approach to understanding celestial and terrestrial bodies and motion. He granted that one may, therefore, look with Duhem at the decree as the "starting point of a new era in scientific thinking, provided it is kept in mind that the decree expressed rather than produced that climate of thought."[303]

Jean Buridan

However strong and spectacular the stream of reaction to Duhem's 1913 first volume *Le système du monde: histoire des doctrines cosmologiques de Platon à Copernic* (*The System of World: A*

---

[301] Grant, *Nature of Natural Philosophy*, vii.

[302] Grant, *Nature of Natural Philosophy*, vii; Jaki, *Science and Creation*, 231.

[303] Jaki, *Science and Creation*, 230.

History of Cosmological Doctrines from Plato to Copernicus), his work has provided undeniable evidence that medieval faith in the predictability of nature was rooted in the theology of the "Maker of Heaven and Earth."[304] It was not just a single belief, but a climate of shared belief nurtured by an educational system comprised of universities, cathedral schools, and monasteries that consistently taught Christian theology. Circumnavigate the conclusion however one may, the theological beliefs that united the consistent learning centers teaching those beliefs *did not exist* in any of the ancient cultures in which science was stillborn.

Thus far, several histories have been summarized. First, the history, meaning, and definition of "science" as the quantitative analysis of objects in motion, was explained. Second, the history of ancient cultures and their skills, masteries, and failures (stillbirths) to produce a viable science of physical laws and systems of laws, all of them based on a pantheistic worldview, was presented. Third, the history of the Old Testament cultures and the worldview instilled by the theology of a personal and merciful God who created a universe of order was described. Fourth, the history of Christianity, from the first millennium and into the second millennium when the Greek works were introduced and translated, was reviewed. The history of scholars who dared to reject certain long-held ideas because they contradicted the tenets of divine revelation shows the significance of the difference in the Christian worldview and the pantheistic worldview. It is within the histories presented thus far that Jaki named the "classical and most influential case" that represents the birth of modern science from Christianity, Jean Buridan (1300–1358), the French priest who developed the concept of the *impetus* which led to the modern concept of inertia and paved the way for Isaac Newton's first law of motion.

---

[304] Jaki, *Science and Creation*, 231.

In his work *Quaestiones super quattuor libris de Cælo et Mundo*, Buridan showed that a radical departure from Aristotelian cosmology and physics was *absolutely necessary* for explaining the movement of bodies. Buridan not only departed from untenable ideas, he affirmed his faith in the Creator and derived from those "articles of faith" that could only be known by revelation and not by scientific demonstration. Buridan stated that "in many an instance one should not believe Aristotle who made many propositions contrary to the Catholic faith because he wanted to state nothing except what could be derived from considerations based on what is seen and experienced."[305] Stated more concisely, and this should be considered carefully–it was faith, not observation, experimentation, or investigation that gave the first breaths to modern science.

Buridan's theory of *impetus* is found in Book VIII, Question 12 of *Super octo libros physicorum Aristotelis subtilissimae quaestiones*. He was thinking about what moves a projectile after it leaves the hand of the projector. It is first necessary to understand what Aristotle asserted, which was the accepted explanation in Buridan's day, so first a brief review.

Aristotelian theory of motion held that terrestrial bodies had a natural motion towards the center of the universe, which meant the center of earth. Motion in any other direction was "violent" motion because it contradicted natural motion and thus, required a mover to move it. Bodies were thought to naturally desire rest, so whenever something moved in any other way than naturally, there had to be a mover in contact with it.[306] If the mover ceased to move it, the body fell straight to the earth and became suddenly at

---

[305] Jaki, *Science and Creation*, 232; quoting *Quaestiones super libris quattuor de cælo et mundo* edited by E. A. Moody (Cambridge, MA: The Medieval Academy of America, 1942), 152 (Lib. II, quaest. 6).

[306] Herbert Butterfield, *The Origins of Modern Science* (New York: The Free Press, A Division of Simon & Schuster Inc., 1957) 15-16.

bodies, which in turn meant he had to ponder the cause of motion for terrestrial bodies, and he did so in the same atmosphere in which the Condemnations of 1277 were made.

So, in Book VIII, Question 12 of the above mentioned work, Buridan appealed to common experience and judged Aristotle's position to be unsatisfactorily solved.[309] He gave the example of a child's toy, the top. When a top spins, it spins in place so there is no vacuum left behind and thus no *antiperistatic* effect to impel the top to keep spinning. As a second example, he described the "smith's wheel" and how it also moves in a circular motion but does not leave a vacuum. As a third example, he pointed out that if an arrow were sharp at both ends, it would still move in the same way as it would move if the back end were blunt. If the motion were caused by the impulsion of the air moving in behind the arrow as it pierced the air, the arrow with a sharp posterior should not fly as far, but this is not observed.

As a fourth example, he described the scenario of a ship moving through water. If the ship is going against the flow and the rowing is stopped, the ship continues on for a while and does not stop immediately. A sailor on deck, however, does not feel the air behind him pushing (impelling) him. He instead feels only the air in front of him resisting him. And if the man were standing at the back of the ship, the strong force from the air rushing in behind the ship and pushing it along ought to knock the man violently into the cargo. Experience shows in all of these scenarios that *antiperistasis* is false.

---

[309] This question can be found translated online at Professor Gyula Klima's website, Philosophy, Fordham University at http://faculty.fordham.edu/klima/Blackwell-proofs/MP_C23.pdf, a reproduction of pages 532 to 538 of Marshall Clagett, *The Science of Mechanics in the Middle Ages* (Madison, WI, The University of Wisconsin Press, 1959).

Buridan then argued that if, fundamentally, motion is maintained by continuous contact with a mover, then there is no explanation for how the top or the smith's wheel can continue to move after the hand is removed, for even if a cloth surrounds the top or the wheel on all sides blocking any movement of air, it still spins after the hand is removed. Further, he argued, common experience shows that when a person pushes his hand through the air, he does not feel the air behind his hand pushing it along whether he has a stone in it or not. Buridan concluded that since, in those cases, there is no air to impel motion, no hand to sustain it, no rowing to move it, there must be another explanation. This is how he arrived at his *impetus* theory:

> Thus we can and ought to say that in the stone or other projectile there is impressed something which is the motive force (*virtus motiva*) of that projectile. And this is evidently better than falling back on the statement that the air continues to move that projectile. For the air appears rather to resist. Therefore, it seems to me that it ought to be said that the motor in moving a moving body impresses (*imprimit*) in it a certain *impetus* or a certain motive force (*vis motiva*) of the moving body, [which *impetus* acts] in the direction toward which the mover was moving the moving body, either up or down, or laterally, or circularly. And by the amount the motor moves that moving body more swiftly, by the same amount it will impress in it a stronger *impetus*.[310]

The *impetus* continues to move a stone after the hand throws it, and the *impetus* is continually decreased by the resisting air

---

[310] Jean Buridan, Book VIII, Question 12 of *Super octo libros physicorum Aristotelis subtilissimae quaestiones* at http://faculty.fordham.edu/klima/Blackwell-proofs/MP_C23.pdf, paragraph 4.

and by the gravity of the stone. He also related *impetus* to mass:

> Hence by the amount more there is of matter, by that amount can the body receive more of that *impetus* and more intensely (*intensius*). Now in a dense and heavy body, other things being equal, there is more of prime matter than in a rare and light one. Hence a dense and heavy body receives more of that *impetus* and more intensely, just as iron can receive more calidity than wood or water of the same quantity. Moreover, a feather receives such an *impetus* so weakly (*remisse*) that such an *impetus* is immediately destroyed by the resisting air. And so also if light wood and heavy iron of the same volume and of the same shape are moved equally fast by a projector, the iron will be moved farther because there is impressed in it a more intense *impetus*, which is not so quickly corrupted as the lesser *impetus* would be corrupted. This also is the reason why it is more difficult to bring to rest a large smith's mill which is moving swiftly than a small one, evidently because in the large one, other things being equal, there is more *impetus*.[311]

Tying this reasoning to common experience, Buridan also explained that this is why one who wishes to jump a longer distance takes a few steps back to run faster and drive himself farther, and why the jumper does not feel the air propelling him but rather the air in front of him resisting him against the force of his jump.

Finally, Buridan turned this path of reasoning toward the heavens and noted that the Bible does not claim that God *had* to keep his hand on the celestial bodies to maintain their

---

[311] Buridan, *Super octo*, paragraph 6.

motion.[312] Buridan suggested that the motion of celestial bodies could be answered another way.

> God, when He created the world, moved each of the celestial bodies as He pleased, and in moving them He impressed in them *impetuses* which moved them without His having to move them any more except by the method of general influence whereby He concurs as a co-agent in all things which take place; "for thus on the seventh day He rested from all work which He had executed by committing to others the actions and the passions in turn." And these *impetuses* which He impressed in the celestial bodies were not decreased nor corrupted afterwards, because there was no inclination of the celestial bodies for other movements. Nor was there resistance which would be corruptive or repressive of that *impetus*.[313]

In other words, Buridan introduced the concepts that would lead to Newton's first law of motion, that a body at rest would stay at rest and a body in motion would stay in motion with the same speed and in the same direction unless acted upon by another force. The pantheistic worldview never would have led to such an idea because it was fundamentally and institutionally opposed to it. Buridan's insight derived from his faith in the Christian Creed, divine revelation applied to reason and observation, which led to exact science as a self-sustaining enterprise of physical laws and systems of laws. "*I might seek from the theological masters what they might teach me in these matters as to how these things take place.*"

Therefore, following the Condemnations of 1277 by Tempier against a set of tenets upheld by Aristotle and his followers, a large movement appeared that liberated Christian

---

[312] Buridan, *Super octo*, paragraph 6.
[313] Buridan, *Super octo*, paragraph 6.

thought from the ancient Greek thought and produced modern science.[314] Duhem is considered to have identified the 1277 articles as the most significant event in the birth of modern science, while Jaki highlighted the spark ignited by Buridan a generation later.[315] For Jaki, however, it is not a certain man, event, or date that marks the birth of science though; it is a breakthrough in a naturalistic worldview that rejected the pantheistic doctrine of eternal cycles and approached the investigation of nature guided by the light of Christian faith in a merciful, faithful God who created the world out of nothing with an absolute beginning and end in time, that is ordered, predictable, and stable, but also not a god itself. This *breakthrough*, based not on observation or experiment but on divine revelation and faith, is thus the birth of modern science, a fundamental departure from the worldviews in which modern science was stillborn.

---

[314] Pierre Duhem, *Essays in the History and Philosophy of Science*, Translated and Edited by Roger Ariew and Peter Barker (Indianapolis, IN: Hackett Publishing Company, 1996), 244.

[315] Jason Gooch, "The Effects of the Condemnation of 1277," *The Hilltop Review*, Vol. 2 (2006), 34; quoting Duhem, Pierre, *Études sur Leonard de Vinci*, Vol. I (Paris: Hermann, 1906), 412.

Chapter 3 - "Of Christianity"

Chapter 3 - "Of Christianity"

Jaki referred to Buridan's essay on *impetus* theory as the "most important passage ever penned in Western intellectual history as far as science is concerned."[1] This statement can be misunderstood if the context, thus given in this book, is not grasped. Jaki did not identify the passage itself as the birth of science, but Buridan's rejection of Aristotelian eternal cycles that was drawn from the Old Testament and Christian tradition—the view of creation by a Creator with a beginning in time and a rational, ordered, and stable cosmos. This "birth" was a conception that emerged in the mind of humanity, theologically similar to the generation of a thought in God, the procession of the Word, the Son who was incarnate of the Virgin.

Buridan was of course not the first Christian to oppose Aristotle on the eternity of the world. The Fourth Lateran Council in 1215 codified the dogma that there is "only one true God, eternal and immeasurable, almighty, unchangeable, incomprehensible and ineffable, Father, Son and holy Spirit, three persons but one absolutely simple essence, substance or nature . . . one principle of all things, creator of all things invisible and visible."[2] The Condemnations of 1277 in the generation that preceded Buridan identified Aristotle's errors that were contrary to the Christian dogma of creation out of nothing and with a beginning in time, and to the common experience of objects in motion.

Buridan was not a theologian, but a man with a brilliant scientific mind. While confident in his faith to guide his thinking and lay boundaries for reality, he was most interested in explaining natural phenomenon, particularly the motion of objects, and even more particularly the beginning of all

---

[1] Jaki, *A Late Awakening*, 49.
[2] Fourth Lateran Council, 1215, paragraph 1, see *DS*, 800.

motion. His assent in faith to the tenets of the Christian Creed guided him to assert the most critical breakthrough in the history of science, the idea of inertial motion and *impetus*.[3]

This idea could also be called the beginning of physics, exact science, where one discovery generates the next at an ever more accelerated rate. Buridan became the rector of the University of Paris in 1327 and taught there until about 1360. In 1377, his theory was formally proposed by Nicole Oresme (1320–1325) and was destined to be adopted by Albert of Saxony (1316–1390), Nicolaus Copernicus (1473–1543), Galileo Galilei (1564–1642), and Sir Isaac Newton (1642–1727). As has, in fact, occurred since the time of Buridan, physics has grown exponentially with new insights, understandings, capabilities, and realms of observation and measurement at almost unimaginable scales of minuteness and grandeur.

Buridan's *impetus* theory did not occur to all of those scholars before him who read Aristotle's *Physics* and *De cælo*, scholars who actually had the same interest as Buridan in explaining the motion of objects. There were, of course, medieval Christian scholars who did not arrive at Buridan's conclusions either, which is why it is better not to say that Buridan himself gave birth to modern science, but the *mentality* that arose from the Christian tenets. That is not to diminish Buridan's genius, but to admit that there were also Jewish, Muslim, and other scholars as capable as Buridan. They, however, operated under a radically different worldview. Arguably, they never would have arrived at Buridan's conclusions about the beginning of motion because they fundamentally held beliefs that were incompatible with such a view of the cosmos.

One may wonder how the Jewish or Muslim monotheism is any different from the Christian monotheism. The Jewish and Arab scholars held a monotheistic belief that God

---

[3] Jaki, *A Late Awakening*, 50.

created everything in the beginning, just as Christian scholars did. Why were they unable to break from Aristotle's pantheistic doctrine of motion based on the doctrine of eternal cycles and returns, then? This question is valid and one that Jaki addressed.

The short answer is that there was too great a separation of philosophy from theology. This idea of eternal cycles had, as Jaki put it, a "stranglehold" on ancient cosmology.[4] Even though the Greeks viewed the universe as having a beginning, it was not an absolute beginning like the absolute beginning in time in the Bible. According to Aristotle and the ancient Greeks, the universe emanated from the First Cause which was eternal, not unlike the Christian theology of Creator and creation. The meaning of "emanate" is important here. In translation, the Latin word is *ē*, meaning "out from," and *mānāre*, meaning "to flow." To say the universe emanated from the First Cause or God is to say it flows out, which is vastly different than saying the universe was created.

In this emanation, there was a cycle with a beginning, but not an absolute beginning. The beginning was only the mark of a new cycle, a cycle that repeats eternally. For the Greeks, a cycle was called the Great Year and lasted, they thought, 36,000 years. For the other major religions, such as those of the Chinese and Babylonians, the cycle time was different. The monotheism of the Jews and the Muslims perhaps too easily accommodated by this concept of emanation, but Christian monotheism could not at all because of the Incarnation and the Holy Trinity. This difference requires an explanation and emphasis.

First, the distinction between a Creator and creation is most important. It is the distinction between the *act* of Creation out of nothing and natural *change*, the processes that occur in already created things. Creation out of nothing is not a process with a beginning, middle, and end. It is a simple

---

[4] Jaki, *A Late Awakening*, 52.

reality of all existence completely dependent on God.[5] This distinction was not, however, invented by Buridan; it was clarified in the writing of Aquinas, who was canonized by the Church in 1322 when Buridan was a young man.[6] This distinction between the act of Creation *ex nihilo* and causation and the process of change can be found in other theological sources, including Tertullian (156–230) and St. Augustine (354–430) in the early Church. It was also prevalent in the Old Testament, where it was first mentioned. This distinction is also *not* found in the writings of Jewish and Arabic medieval scholars since they were less ready to see that an eternal universe contradicted Scripture.

Second, this distinction between the Creator and creation is owed to the unique Christian monotheism, which is a Trinitarian and Incarnational monotheism unlike any other. There is a major difference in Christian monotheism and Jewish or Muslim monotheism, and to grasp the significance of Buridan's breakthrough in the history of science, the Trinitarian and Incarnational aspect of Christianity must also be understood. The revelation of the birth, life, death, and resurrection of Christ taught the reality of the Trinitarian nature of God and the divinity of Christ. God is one God in three Persons, the Father, the Son, and the Holy Spirit. Christ is the Second Person of the Holy Trinity, the Son, who became man. Therefore, *Christ is also the Creator.* Christ is also called the Word and the *Logos,* Rationality Itself, which explains why there is order and predictability in physical laws created by this personal and merciful God.

---

[5] A good explanation of the Cosmogonical Fallacy which still plagues the understanding of evolution and intelligent design was written by Michael W. Tkacz, "Aquinas vs. Intelligent Design" Catholic Answers Magazine, Volume 19, Number 9 (November, 2008) at http://www.catholic.com/magazine/issues/volume-19-number-9.

[6] Jaki, *A Late Awakening,* 55.

The Jewish and Muslim faiths do not acknowledge Christ or the hand of God in salvation history. Without the dogma of the Holy Trinity and the Incarnation, there is not much to save monotheism from the errors of pantheism with an eternal and cycling universe emanating from a First Cause because the contradiction with revelation is not as clear. There is a certain scientific significance in the beginning of St. John's gospel:

> At the beginning of time the Word already was; and God had the Word abiding with him, and the Word was God. He abode, at the beginning of time, with God. It was through him that all things came into being, and without him came nothing that has come to be. In him there was life, and that life was the light of men. And the light shines in darkness, a darkness which was not able to master it.[7]

John also referred to Christ as the "Word made flesh" and the "Father's only-begotten Son full of grace and truth."[8] John writes, "And the Word was made flesh, and came to dwell among us; and we had sight of his glory, glory such as belongs to the Father's only-begotten Son, full of grace and truth." (*Et Verbum caro factum est, et habitavit in nobis: et vidimus gloriam ejus, gloriam quasi unigeniti a Patre plenum gratiæ et veritatis.*) John used the Greek words μονογενοῦς παρὰ, translated as *monogenes*, the "only begotten" Son of the Father, and in Latin as *unigeniti*.

Jaki developed this concept in several of his essays. In ancient Greece the word *monogenes* referred to the eternally emanating cosmos, *unigenitus*, also *universum* or universe.[9] To Plato, for instance, the *monogenes* was the Unknown God, the

---

7 Knox, John 1:1-5.
8 Knox, John 1:14.
9 Jaki, *A Late Awakening*, 55.

cosmos itself. Plotinus, six hundred years after Plato, still referred to the *monogenes* as the Unknown God. When John called Christ by the same words, it marked a radically different view of God: the Unknown God was named the Christian God, a Trinitarian and Incarnational God. The god of the pantheists in ancient Greece is drastically different from the God of the Gospels. If this theological point is missed by historians, the rationality of the Greeks will not appear all that different from the rationality of the Christians, but to miss that point is to miss the mindset, the worldview, the radically different psychology of Christianity.

This theological point is connected with Jaki's description of "science" and "religion" as entities separated by God Himself, as discussed at the end of the chapter on the definition of "science." Since science deals with quantities and measurement of objects in motion and religion deals with the ultimate purpose of mankind, Jaki also noted that Jesus indirectly warned that science should be kept in its secondary place. When Jesus taught his followers to "Seek first the Kingdom of God and all else will be given to you," he was telling them that His Kingdom is supernatural, not natural, just as he told Pilate, "My Kingdom is not of this world."[10] Though not usually used as such, those words from Our Lord offer excellent advice to scientists.

The fruits of Christ's divinity certainly have been relevant for this world, including the scientific aspect of history, but the main lesson of that divinity is that it points to a world beyond this world, beyond the cosmos, beyond the universe. Why is that lesson important when considering science or the history of science? The lesson is important because it is a reminder that the questions religion can answer are far more important questions to humanity than the questions science can answer. If one is in agreement, then, with the late Fr. Stanley Jaki, it can be asserted with demonstrated confidence

---

[10] Knox, Matthew 6:33 and John 18:36.

that for natural sciences to be born, supernatural revelation was needed. *"There had to come a birth, the birth of the only begotten Son of the Father as a man, to allow science to have its first viable birth."*[11]

---

[11] Jaki, *A Late Awakening*, 60.

Chapter 4 - Critics

Chapter 4 - Critics

In the telling of things, debates still oscillate about the meaning of the word "science," which was clarified before. Chances are, when someone is first offered the definition of science as *"exact* science, the *quantitative* study of the quantitative aspects of *objects* in motion" there will be a reaction somewhat accusative of *naiveté*. As exalted as science has become, that definition seems inadequate. Chances are, when that same someone is asked for a better definition of science, the response will be to offer a longer definition of science that says the same thing. For example, someone could offer the Merriam-Webster definition, "knowledge about or study of the natural world based on facts learned through experiments and observation."

Of course the defender of Jaki's definition will need to be able to explain why those two definitions are the same. Here is why: The natural world is not the supernatural world. Therefore, the "natural world" is limited to physical things, i.e. objects. To "study" and "learn facts" about objects through "experimentation and observation," one must make measurements of the objects' motion. Since nothing observable or subject to experiment exists in a state of absolute immobility, the only knowledge that can be gained from science, per the definition given, is *"exact* science, the *quantitative* study of the quantitative aspects of *objects* in motion." If the interlocutor insists that science is more than that, a discussion of the immaterial realm necessarily must ensue, which leaves the arena of science and enters the one of reasoned discourse. If this defense is practiced enough, the wisdom of its conciseness becomes clear. Science is limited.

Stephen Barr, theoretical physicist and cosmologist, wrote in an article about Jaki's death that Jaki was known to quote with approval Duhem's statement that "in order to speak of questions where science and Catholic theology touch one another, one must have done ten or fifteen years of study in the pure sciences."[1] This ambiguity about the definitions of science and theology seems to be the point of deviation for Jaki's critics who were not scientists or theologians *in addition to* being historians. It seems that to effectively delve into the "searching questions" that Jaki explores in the history of science, questions that are tied to the history of theology, being a historian is secondary or tertiary to being a scientist and a theologian. Certainly that is debatable, but of course, that is the order of Jaki's scholarly training, so even if someone disagrees that a historian of science is benefitted by first being a scientist and a theologian, it is enough to at least grasp that for Jaki, this was the case.

At any rate, there are still livelier debates that swing to and fro about the genesis, beginning, origin, and foundations of modern science, which the longest chapter of this book sought to crystallize by searching with Jaki back through history. As evidenced by the volumes still being printed about history of science, there is a significant lack of appreciation for the literal impossibility of defining a "beginning," a "genesis," an "origin," or the "foundation" of science *if* the concept of science is not defined at the outset. Jaki understood the need to clarify both, and where other historians tried to pinpoint beginnings amid ambiguity, Jaki clarified the definition of science so he could search through history and explain the birth of science. It is one thing to study the history and make note of how scientific thinking evolved over time in different cultures. It is quite another to ask why and how breakthroughs in understanding were made. Both approaches are beneficial, but they have different purposes.

---

[1] Stephen M. Barr, "Science Seeking Understanding," *First Things* (June/July 2009) at
http://www.firstthings.com/article/2009/05/science-seeking-understanding-1243195356.

Jaki used the term "birth" of science intentionally because he argued that the "Savior of Science" is the Savior of Mankind; therefore, a "Saving Birth" was needed for modern science to be born because it oriented the way Christians view the world and nature. Of course science is a human endeavor, but since man is made in the image of God with intellect and free will, science without faith in revealed religion will not have the light of grace to guide it.

A "birth" also implies more than just a moment in time. Births do not happen as a single instant devoid of any other occurrences. Gestation occurs over a long period of time, and survival to birth is not a guarantee. For such survival, the womb must be protective and nurturing. For science, that gestation and nurturing came from the biblical worldview. Whatever is "born" also remains dependent following the emergence from the womb for at least the length of time the gestation lasted, followed by a slow weaning to independence but never a full division in the relationship. The ideas in this paragraph are not explicitly expressed by Jaki, but the "birth" of science is a theological concept that could be expanded to demonstrate why the Holy Mother Church must play a role in guiding scientific endeavors for the future of mankind.

The work of Pierre Duhem in the early twentieth century brought to light the role the Church had in the birth of modern science. More and more historians agree that not only did modern science have a beginning, it had a beginning in the Christian West and not accidentally, but because the Christian psychology was conducive to a naturalistic, realistic view of the universe. Jaki was an esteemed and awarded historian of science, physicist, and theologian, sought after for talks and publications, but the reactions to his rather stark claims have been mixed. Jaki was the first to consider systematically the "stillbirths" of science in all major ancient cultures. He was also unique, during his time, in emphasizing the theological aspect of the history of science and in demonstrating that the birth of modern science was not an accident of Christianity, but a *conscious* conclusion.

In his book (mentioned in the introduction), *Creation and Scientific Creativity: A Study in the Thought of S. L. Jaki*, first published in 1991 and updated in a second reprint in 2009, the year of Jaki's death, Paul Haffner responded to Jaki's critics in a full chapter. The first part of the chapter deals with the particular historical argument. Although Jaki's detailed research on the stillbirths of science exceeded the extent of the research by Alfred North Whitehead, they both agree on the contribution of medieval scholasticism to the unique birth of science in Europe.

In *Science and Creation*, Jaki quoted Whitehead in a long passage where Whitehead, who considered himself agnostic but turned to religion later in life although he never joined any specific institution, acknowledged the climate of thought as the most crucial ingredient for the eventual breakthrough of modern science.[2]

> I do not think, however, that I have even yet brought out the greatest contribution of medievalism to the formation of the scientific movement. I mean the inexpugnable belief that every detailed occurrence can be correlated with its antecedents in a perfectly definite manner, exemplifying general principles. Without this belief the incredible labours of scientists would be without hope. It is this instinctive conviction, vividly poised before the imagination, which is the motive power of research: that there is a secret, a secret which can be unveiled. How has this conviction been so vividly implanted on the European mind?

---

[2] Jaki, *Science and Creation*, 230; Whitehead, *Science and the Modern World*, 17-18; "Alfred North Whitehead – Biography," The European Graduate School website at http://www.egs.edu/library/alfred-north-whitehead/biography/.

When we compare this tone of thought in Europe with the attitude of other civilisations when left to themselves, there seems but one source for its origin. It must come from the medieval insistence on the rationality of God, conceived as with the personal energy of Jehovah and with the rationality of a Greek philosopher. Every detail was supervised and ordered: the search into nature could only result in the vindication of the faith in rationality. Remember that I am not talking of the explicit beliefs of a few individuals. What I mean is the impress on the European mind arising from the unquestioned faith of centuries. By this I mean the instinctive tone of thought and not a mere creed of words.

In Asia, the conceptions of God were of a being who was either too arbitrary or too impersonal for such ideas to have much effect on instinctive habits of mind. Any definite occurrence might be due to the fiat of an irrational despot, or might issue from some impersonal, inscrutable origin of things. There was not the same confidence as in the intelligible rationality of a personal being. I am not arguing that the European trust in the scrutability of nature was logically justified even by its own theology. My only point is to understand how it arose. My explanation is that the faith in the possibility to science, generated antecedently to the development of modern scientific theory, is an unconscious derivative from medieval theology.[3]

Whitehead called it an "unconscious derivative," but Jaki, as said, argued that it was quite conscious. It does seem more reasonable to assume that the medieval Christian scholars were consciously aware of their faith than to assume they did

---

[3] Whitehead, 13-14.

not realize it guided them. Perhaps they did not realize how much it guided them, though.

Haffner also noted, as Jaki did, that there has been an "almost systematic oversight of Duhem and of the Middle Ages in books written during the last two or three decades by Protestant scholars on the rise of science and Christianity."[4] From M. B. Foster in the 1930's to Langdon Gilkey in the 1950's and 1960's to Reijer Hooykaas, Donald MacKay, and Eugene Klaaren in the 1970's to G. B. Deason in the 1980's, one would "look in vain" for an appreciation of either Duhem's or Jaki's research.[5] One exception is in the work of Herbert Butterfield. In his 1957 book, *The Origins of Modern Science*, he praised the work of Duhem for bringing out the importance of the fourteenth-century teaching on the subject of *impetus*, and even began his book with a chapter about it.[6]

In 1992, David C. Lindberg, who described himself as a "liberal Protestant," also mentioned Duhem rather extensively in a discussion about the Condemnations of 1277 in his book *The Beginnings of Western Science: The European Scientific Tradition in Philosophical, Religious, and Institutional Context, Prehistory to A. D. 1450.*[7] In his consideration of the Condemnations of 1277 and Duhem's interpretation of them, he wrote:

> Pierre Duhem, writing early in the twentieth century, saw the condemnation of 1277 as an attack on entrenched Aristotelianism, especially Aristotelian physics, and therefore as the birth certificate of modern science. [. . .] But to place the emphasis here is to miss the primary

---

[4] Haffner (2009), 145.

[5] Haffner (2009), 145.

[6] Butterfield, *The Origins of Modern Science*, 27.

[7] Bernard O'Connor, "The Catholic Church and Galileo," (2009) EWTN website at
http://www.ewtn.com/library/ISSUES/churchgalileo.htm.

significance of the condemnations. Duhem viewed the condemnations as the key event in the shattering of Aristotelian orthodoxy, but in 1277 no such orthodoxy existed; the boundaries and the power relationship between Aristotelian philosophy and Christian theology were still being negotiated, and the degree to which Aristotelianism would acquire the status of orthodoxy was not yet clear.[8]

As said already, Jaki noted in his 1974 book *Science and Creation* that one may therefore look with Duhem at the decree of 1277 as the "starting point of a new era in scientific thinking, provided it is kept in mind that the decree expressed rather than produced that climate of thought."[9] If any "orthodoxy" existed, it was Christian orthodoxy, not Aristotelian, and Duhem knew this. It is doubtful that after writing the ten volumes of his *Système du monde* Duhem thought that Aristotle's teaching was orthodox for Christians. Aristotelian thought *was* orthodox to Greeks and was accepted by the Muslims, so any Christian philosopher who studied the Arab and Greek texts had to deal with the entire system and not merely dismiss what contradicted Christianity without explanation. Obviously this systematic examination took time, so the delineations were not immediately clear. Lindberg referenced Duhem's *Études sur Léonard de Vinci* to support the claim that Duhem thought the condemnations shattered the Aristotelian orthodoxy, but that does not seem to be what Duhem meant. In that reference, Duhem actually wrote: "The Christian orthodoxy therefore required, it seems, a waiver of various principles of Aristotelian physics. [. . .] If

---

[8] Lindberg, Kindle location 4425-4431.
[9] Jaki, *Science and Creation*, 230.

we were to assign a date to the birth of modern science, we would probably choose 1277."[10]

Duhem certainly did not seem to suggest that Aristotelian thought was "orthodox" before the Condemnations of 1277. Nonetheless, Lindberg was clear that he disagreed with Duhem.

Lindberg viewed the Condemnations of 1277 more as a confrontation between "liberal and radical efforts to extend the reach and secure the autonomy of philosophy" and a "conservative backlash" of a "sizable and influential group of traditionalists not yet ready to accept the brave new world proposed by the . . . Aristotelians."[11] In his opinion, "to put the event in its proper light, the condemnations represent a victory not for modern science but for conservative thirteenth-century theology."[12] He saw the condemnations as a "ringing declaration of the subordination of philosophy to theology."[13] His commentary gives the impression he viewed the Condemnations of 1277 negatively, but considering that this subordination purified and Christianized Aristotelian physics of its pantheism, that "ringing declaration" can also be viewed positively for the development of modern science. At the very least, it was recognition of the limits of philosophy.

Even though his book title indicated that it was about the "beginnings" of Western science, Lindberg did not actually pinpoint a beginning, but several revolutions that may have led to the beginning. Neither did Lindberg clearly define science. In the first chapter of his book, he wrote that "we have no choice but to accept a diverse set of meanings as

---

[10] Pierre Duhem, *Études sur Léonard de Vinci* (Paris: A. Hermann, 1906), 412.
[11] Lindberg, 248.
[12] Lindberg, 248.
[13] Lindberg, 248.

legitimate and do our best to determine from the context of usage what the term 'science' means on any specific occasion."[14] This is not to criticize his approach, for it is the masterful work of a master historian, but to contrast it with Jaki's approach and to defend Jaki's approach as the one that offers more insight into how science can be advanced in the future. No historian can be absolutely objective in interpreting historical times. (A scientist might even say that it is easier to succumb to bias in the study of historical data than in the assessment of scientific data because history deals with human persons with free will as data points. Quantitative data has no volition of its own; the numbers speak for themselves.) At least for Jaki, who was a theologian, there is an admitted lens through which he explored the history of science—he studied the history of theology along with it to see how the two affected each other in different cultures.

While both historians delved into the scientific beginnings in ancient cultures, Jaki began by defining what science *is* and always should have been even before the importance of applying mathematics to observations in nature was finally realized, and he started asking questions about *why* that importance failed to be realized in some cultures and was ultimately realized in the Christian West. One of Lindberg's ending conclusions was that Alexandre Koyré "put his finger in the right place" regarding the underlying source of revolutionary novelty in the sixteenth and seventeenth centuries, a metaphysical and cosmological revolution toward a view of nature that was *mechanical* rather than animistic, which Jaki would have agreed is partially true. However, Koyré rejected the century of experimentation after Buridan and Oresme as the real beginning of a new scientific outlook, which is a direct contradiction of Jaki's conclusion.

Koyré also had a disagreement with another historian, Alistair Cameron Crombie. Lindberg wrote of both

---

[14] Lindberg, 2.

historians, noting that they traded opinions in the 1950's and 1960's. Crombie argued, against Koyré, that "a systematic theory of experimental science was understood by enough [thirteenth- and fourteenth-century] philosophers . . . to produce the methodological revolution to which modern science owes its origins." For Crombie it was the beginning of experimentation that ushered in the Scientific Revolution. While Koyré and Crombie disagreed on the beginning of a new scientific outlook, Jaki disagreed with them both. He did not see the new outlook as a result of either a new view of nature or experimentation, but of something more fundamental.

Jaki's reply in *Science and Creation* was that Koyré did not consider in sufficient detail that the new mechanical view of nature had a great deal to do with the theological rejection of Greek pantheism.[15] Historians of science, as respecters of facts, are not expected to show much sympathy for theological underpinnings, but even so Jaki insisted on a broader view of the facts and he asserted one large fact that cannot be denied by anyone studying the history of science, the "collective faith of the Middle Ages is a *fact* of history, and so is its enormous impact on the modern mind so proud of its science."[16] The emphasis was Jaki's.

Jaki also wrote in his 2002 autobiography that he was aware that the suggestion that science was owed substantially to Christianity, indeed to Christ's birth, would exclude him from certain scholarly circles. He noted that Crombie, being a fellow Catholic, could have lent his theological work more support, but Crombie "chose to look the other way."[17] The relationship between the two men is perplexing. Crombie came to most of Jaki's Fremantle lectures in 1976, the lectures upon which he based the book *The Origin of Science and*

---

[15] Jaki, *Science and Creation*, 235.
[16] Jaki, *Science and Creation*, 235.
[17] Jaki, *A Mind's Matter*, 62.

the Science of its Origin, but since Crombie had slighted Duhem in his much earlier 1959 book, *Medieval and Early Modern Science*, Jaki did not expect Crombie to "take much delight in [his] thesis," according to his autobiography.[18]

Jaki and Crombie were both members of the Pontifical Academy of Sciences. Crombie was elected in 1994, four years after Jaki was appointed in 1990 as an honorary member by Pope John Paul II. Jaki even gave a lecture about science and culture in 1994 at the Pontifical Academy when Crombie was elected. The two men met, but for whatever reason, they never agreed. Their meeting at the Pontifical Academy in 1994 was their last meeting, and Jaki wrote in his autobiography that Crombie's position that Greek rationality was the origin of science remained unchanged.[19] For reasons given in the bulk of this book, Jaki could never have agreed with that decidedly un-theological conclusion. Crombie published a major work that same year, three thick volumes of entitled *Style of Scientific Thinking in the European Tradition*. Of that work, among other strong criticism, Jaki wrote that Crombie's few references to Duhem "remain a classic in disproportionality."[20]

The historian Edward Grant, on the other hand, referenced both Duhem and Jaki considerably in his books, though he puts more emphasis on the Catholic learning institutions than on Catholic theology. Either way, the two are hardly separable. Grant prefaces his 2010 book, *The Nature of Natural Philosophy in the Late Middle Ages*, with a quote from another of his books from 1996, *The Foundations of Modern Science in the Middle Ages*:

I would argue that in the Latin Middle Ages of Western Europe an intellectual environment was established that

---

[18] Jaki, *A Mind's Matter*, 62.
[19] Jaki, *A Mind's Matter*, 63.
[20] Jaki, *A Mind's Matter*, 62.

proved conducive to the emergence of early modern science. The new intellectual environment was generated and shaped by "certain attitudes and institutions that were generated in Western society from approximately 1175 to 1500. These attitudes and institutions were directed toward learning as a whole and toward science and natural philosophy in particular. Together they coalesced into what may be appropriately called 'the foundations of modern science.' They were new to Europe and unique to the world. Because there is nothing to which we can compare this extraordinary process, no one can say whether it was fast or slow."[21]

Grant seems to credit the institutional apparatus, that is the universities, as the chief harbinger of science.

In 2005 another Catholic historian, political analyst, and *New York Times* best-selling author actually did delve into Jaki's thesis and defend it at length. Thomas E. Woods published a book titled *How the Catholic Church Built Western Civilization*. The book covered a broad range of topics including the contributions of Catholic monks, universities, art, architecture, international law, economics, charities, and morality. The longest chapter in his book was the fifth chapter, "The Church and Science," which he divided into sections dealing with the history of the Galileo Affair, the Cathedral Schools of Chartres and the Condemnations of 1277, the scientists who were also priests (like Jaki), and the scientific achievements of the Jesuits. In an entire ten page section, "God 'Ordered All Things by Measure, Number, Weight,'" Woods presented Jaki's work regarding the history of science and he gave an accurate description of the "stillbirths" of science in other cultures, the influence of the biblical worldview of the Old Testament, the differences between Aristotelian thought and Christian thought, and the

---

[21] Grant, *Nature of Natural Philosophy*, ix.

significance of Buridan's *impetus* theory. He acknowledged Duhem's contributions and praised Jaki as a "prizewinning historian of science–with doctorates in theology and physics– whose scholarship has helped give Catholicism and Scholasticism their due in the development of Western science."[22]

Following in much the same thought as Whitehead, Grant, and Lindberg, another historian, James Hannam, a Catholic convert, published a book in 2011 titled *The Genesis of Science: How the Christian Middle Ages Launched the Scientific Revolution.*[23] Anyone familiar with Jaki's work would expect a mention of it in such a book, yet not a single reference to Jaki can be found in the twenty-one chapters, suggestions for further reading, or extensive bibliography. Duhem is only mentioned once in the introduction:

> The fight back began 100 years ago with the work of a French physicist and historian called Pierre Duhem (1861–1916). While researching an unrelated matter, he came across a vast body of unread medieval manuscripts. What Duhem found in these dusty tomes amazed him. He quickly realized that science in the Middle Ages had been sophisticated, highly regarded, and essential to later developments. His work was carried forward by the American Lynn Thorndike (1882–1965) and the German Anneliese Maier (1905–1971), who refined and expanded it. Today, the doyens of medieval science are Edward Grant and David Lindberg. They have now retired, but their students already occupy exalted places in the universities of North America. As scholars explore more

---

[22] Thomas E. Woods, *How the Catholic Church Built Western Civilization* (Washington DC: Regnery Publishing, Inc., 2005), 75.

[23] James Hannam, *The Genesis of Science: How the Christian Middle Ages Launched the Scientific Revolution* (Washington DC: Regenery Publishing, Inc., 2011).

and more manuscripts, they reveal achievements of the natural philosophers of the Middle Ages that are ever more remarkable.

> Popular opinion, journalistic cliché, and misinformed historians notwithstanding, recent research has shown that the Middle Ages was a period of enormous advances in science, technology, and culture.[24]

This idea that the Christian Middle Ages were the foundations of modern science seems to have taken root among historians of science, but without a real appreciation for the Christian theological roots of it. The trending discourse currently is to acknowledge that modern science developed in the Christian Middle Ages, but to downplay the Christian contribution, or even to call it a myth.

The latter is the case for the Roman Catholic ethicist, Benjamin Wiker whose 2011 book was entitled, *The Catholic Church and Science: Answering the Questions, Exposing the Myths.* Wiker addressed as the "First Confusion" the myth that the Catholic Church is at "war with science" and that "faith is at war with reason."[25] Wiker excellently dispelled the myth as promised with reference to the nineteenth-century historians who promoted the now-exposed, thanks to Duhem, false myth that the Middle Ages were a period of intellectual stagnation. Of John William Draper, who is infamous for his *History of the Conflict between Religion and Science*, Wiker wrote that "although he was certainly an excellent scientist, as an historian of the relationship of the Church to science he was a mere anti-Catholic propagandist."[26] Indeed, posterity seems

---

[24] Hannam, xvi-xvii.

[25] Benjamin Wiker, *The Catholic Church and Science: Answering the Questions Exposing the Myths* (Charlotte, NC: TAN Books, 2011), "The First Confusion."

[26] Wiker, Kindle location 180.

to view Draper this way. Wiker also described how the Church was a generous patron and protector of reason and faith, but then toward the end of this chapter, he added "an additional warning."

> As historian of science Noah Efron smartly states, "For every myth there is an equal and opposite myth." While we might not want to call it a "myth," it is true that a certain kind of reaction to the warfare thesis has occurred among Christians who, tired of the drubbing given to the Faith by the secularists, are zealous to show that Christianity was actually the source, the cause of modern science. It's not difficult to understand how this reaction arose. When historians sifted through the great mound of evidence, reaching way back into the Middle Ages, that had been neglected by the Drapers and Whites, they began to recover the full extent to which Christians had contributed to the rise of modern science. To some of them (especially to irritated Christians), it was as natural to overplay the evidence as it had been for secular materialists to ignore it.

> The general argument goes something like this: "Modern science was born in a particular culture, a Christian culture, and we can trace its antecedents backward all the way into the early Middle Ages. No other culture— Greek, Roman, Indian, Chinese, Egyptian, Babylonian, Islamic, African, Mayan—ever gave us anything like modern science. Therefore, the cause of modern science's success must lie, ultimately, in Christianity, and the more we dig into the Christian origins of modern science, the farther back we find positive evidence for the sustained, sophisticated developments that underlay modern science."[27]

---

[27] Wiker, Kindle location 375-387.

The "general argument" Wiker put in quotes seems to be a paraphrase of Jaki's work, but Jaki's name is not specifically mentioned. The reference for the passage is a chapter written by Noah Efron (mentioned in the beginning of the quote above), "That Christianity Gave Birth to Modern Science." This essay is found in the 2009 compilation of essays, *Galileo Goes to Jail and Other Myths about Science and Religion*, edited by Ronald Numbers who, like Koyré, Needham, Grant, and Lindberg, received the George Sarton Medal for "a lifetime of scholarly achievement" in the history of science.[28] Numbers also is a self-described agnostic, and Noah Efron is a Fellow for the think-tank for Israeli politics, Shaharit.

While a different faith or a lack of faith certainly should not be taken as a lack of credibility as a historian, which would be presumptuous and arrogant, it is fair nonetheless to wonder why the decades of research from an internationally acclaimed Catholic physicist, theologian, historian, author of over fifty works, Templeton Prize winner, and honorary member of the Pontifical Academy of Sciences was not at least mentioned before the essence of his work was dismissed as mythical. Efron quotes Jaki in the opening of his essay, so it is reasonable to assume that Wiker knew the chapter referenced Jaki's work. In Efron's essay the quote from Jaki is the sole reference to Jaki, however. Efron does not present any of Jaki's work in support of the claim that "science was born of Christianity." Instead he addresses the Protestant historians (mentioned before) who systematically ignored Jaki. Then he deems all of it a myth without actually addressing Jaki's research.

---

[28] Ronald L. Numbers, editor, *Galileo Goes to Jail and Other Myths About Science and Religion* (Cambridge, MA and London, England: Harvard University Press, 2009), "Myth 9: That Christianity Gave Birth to Modern Science," Noah J. Efron, 79-89.

Wiker concluded his first chapter by asserting that it is the "position of the Church" to reject this "myth" of the Christian contribution to the birth of science on the doctrinal case that science is a human endeavor because man is made in the image of God, and therefore uses human reason to understand the order of nature.[29] Perhaps Wiker did not intend to dismiss Jaki's work specifically. Perhaps he was unfamiliar with it. Perhaps he did not realize who Jaki was when he referenced Efron's essay with Jaki's work quoted at the beginning. The benefit of the doubt ought to be given, but at the same time a clarification needs to be made. While it is true that the Church would reject a caricature of Jaki's argument that the Church was "*the* source" or "*the* cause" of modern science, it is not true Jaki's argument that science was "born" of Christianity, properly understood, has been rejected by the Church.

After all, Pope St. John Paul II appointed Jaki an honorary member of the Pontifical Academy of Sciences well after the publication of his books putting forth this claim. Pope Benedict XVI honored Jaki in November 2006 while receiving the members of the Academy, of which Jaki was by then an honorary member of sixteen years, by saying, "Fr. Jaki, I thank you for the books you write on science, religion, and creation."[30] Furthermore, in the announcement of Jaki's death, sent to Jaki's fellow Academicians by the President, Nicola Cabibbo, and the Chancellor, Marcelo Sánchez Sorondo, of the Pontifical Academy of Sciences, this aspect of Jaki's work was identified specifically:

The history of science shows that all great creative advances in at least the physical sciences were made in terms of an epistemology which also underlies the

---

[29] Wiker, Kindle location 380.

[30] Jaki, *A Mind's Matter*, "Three More Years" an additional chapter, 17.

classical proofs of the existence of God. These two themes were given a detailed presentation in Prof. Jaki's *Gifford Lectures, The Road of Science and the Ways to God*. Prof. Jaki also believed that historically, this theistic perspective of science emerged from what he called the repeated stillbirths and the only viable birth of science. The former occurred in all great ancient cultures, whereas the latter is intimately tied to medieval Christianity. It was Christianity, and especially its dogma about the divinity of the Incarnate Logos, that gave a special strength to the biblical notion of a coherent universe, fully ordered in all its parts, an idea indispensable to the emergence of Newtonian science. All these themes were set forth in his *Science and Creation* and *The Savior of Science* books.[31]

It is easy to understand why non-Christians would reject a thesis that claimed "science was born of Christianity." It is not so easy to understand why people who profess faith in God the Creator of all things visible and invisible would reject such a claim. Jaki, like Duhem, did not have doctoral degrees in history, but they were, and still are, both recognized as historians of science for the quality and quantity of work—work that indeed changed the received view of the Christian Middle Ages. They were both physicists who made contributions to their fields as well. The most likely reason Jaki's work has not been better appreciated could be that there are biases against accepting his thesis or that it is still not well-understood. The former will not likely resolve, but time and effort can fix the latter.

Jaki will probably be best remembered as a theologian, which is why his insight into the theological history of science

---

[31] Nicola Cabibbo and Marcelo Sánchez Sorondo, Pontifical Academy of Sciences, Circular letter announcing the death of Prof. Stanley Jaki, dated April 8, 2009, available at http://www.sljaki.com/Pontifical_Academy_en.html.

will be valuable. As mentioned in the beginning, Jaki's works take commitment and considerable effort to assimilate. The bulk of his critics, especially the most modern ones, tend not to actually criticize his work, and instead refuted a mischaracterization that has not been researched. As Haffner put it, "any critic of Jaki's position automatically incurs the burden of proof in the debate, a burden hardly ever assumed because of its unusual weight."[32]

Personally, I have spent a great deal of time finding Jaki's sources and verifying his claims as it relates to this aspect of his work, and I found his commentary and conclusions to be accurate, reliable, and insightful. The most difficult task is often in deciphering his subreferences. Entire essays can be written about a single Jaki sentence, so packed with nuance and information are some of them. Reading Jaki's work can be difficult but rewarding. Hopefully this book has presented the reward while easing the difficulty, for it is a sifted compilation of what I found most compelling.

This consideration of Jaki's critics cannot conclude without confronting what seems to be the most negative criticism. In Efron's essay, cited in Wiker's book, offense was obviously taken from the claim that "science was born of Christianity." Efron did acknowledge the "claim that Christianity led to modern science captures something true and important."[33] But then he expressed suspicion:

> When boosters [Jaki and others] insist that "Christianity is not only compatible with science, it created it," they are saying something about science, they are saying something about Christians, and they are saying something about everyone else. About science, they are saying that it comes in only one variety, with a single history, and that centuries of inquiries into nature in

---

[32] Haffner (2009), 147.
[33] Numbers, Efron, 80.

China, India, Africa, the ancient Mediterranean, and so on have no part in that history. About Christians, they are saying that they alone had the intellectual resources-rationality, belief that nature is lawful, confidence in progress, and more-needed to make sense of nature in a systematic and productive way. About everyone else, they are saying that, however admirable their achievements in other realms may be, they lacked these same intellectual resources. Often enough, what these boosters really mean to say, sometimes straight out and sometimes by implication, is that Christianity has given the world greater gifts than any other religion. Frequently, they mean to demonstrate simply that Christianity is a better religion.[34]

[...]

This anything-your-religion-does-mine-can-do-better attitude jiggers one part condescension with two parts self-congratulation, and one wonders why some find it appealing. Yes, Christian belief, practice, and institutions left indelible marks on the history of modern science, but so too did many other factors, including other intellectual traditions and the magnificent wealth of natural knowledge they produced. Assigning credit for science need not be a zero-sum game. It does not diminish Christianity to recognize that non-Christians, too, have a proud place in the history of science.[35]

Although Efron misrepresented Jaki's work almost entirely, his reaction is instructive. The claim that "science is born of Christianity" could sound as chauvinistic to a non-Christian as the claim that "science was born of atheism"

---

[34] Numbers, Efron, 87-88.
[35] Numbers, Efron, 88.

could sound to any believer, and the statement likely will lead critics to criticize without making the effort to understand the full argument. The power of analogy sometimes cuts both ways.

Jaki wrote in an *addendum* to his autobiography in 2009, a few weeks before his death, that he asked himself, "What is the point to work hard on a topic, though only to see in the end that what one tried to transmit on the basis of decades of hard work runs like water off a duck's back?"[36] His disappointment is understandable, but Jaki's hard work has not been wasted. Maybe it ran like water off the back of a duck for some people during his lifetime and maybe there will always be people who do not make the effort to understand his work, but others see Jaki's legacy as a wellspring, an ongoing source of insight and information into the distinctly human activity we call science.

---

[36] Jaki, *A Mind's Matter*, "Three More Years" an additional chapter, 10.

Chapter 5 – What Now?

Chapter 5 – What Now?

In conclusion, it is proposed that if Jaki's thesis is to receive wider acceptance, the way to expand and further his work is to develop a way to communicate it without using language that could seem, from the onset, offensive to other religions. This will require, however, considerable work on the part of the communicator. It will not be enough to agree with Jaki's argument. It will not even be enough to know the essence of it. To communicate the argument, an underlying knowledge of the history and theology must be understood, as well as an ability to defend the definition of science that Jaki used. The effective persuader may summarize and argue only from the surface of the argument, but if the one who seeks to convince does not pull from a depth and breadth of knowledge on the matter, the argument cannot be articulated to adjust to the audience. Any effective educator is aware of the need for flexibility.

Just as Aquinas used the authority of the Old Testament when writing for the Jewish people, the authority of the New Testament when writing for the Christians, and the employment of universal reason when writing for the Muslims, so too should modern evangelists be prepared to tailor their arguments. Some may say the truth is the truth and anyone offended by it should just deal with it, but any parent or teacher knows that such an approach will fail to teach, and most likely will alienate. Such approaches do not consider the human person in the greater context of his or her life in a personal way. Pope Francis recently said in an interview with *America Magazine*, "We must always consider the person."[1] Although he was speaking about homosexuality,

---

[1] Antonio Spadaro, S.J., "A Big Heart Open to God," *America Magazine* (September 30, 2013) at http://www.americamagazine.org/pope-interview.

this seems to be good advice for presenting any argument, especially a potentially contentious one, if the real goal is communication.

The claim that "science was born of Christianity" puts well-meaning non-Christians in a difficult position to either dismiss the research summarily or to risk denying deeply held beliefs possibly rooted in childhood or personal traditions. In other words, perhaps there should be an *ecumenical* consideration, and anyone who knows what ecumenism means also knows it first and foremost does not mean that truth should be compromised. It means truth should be communicated better. There is arguably no truth exempted from the human pursuit of better communication.

For now, Catholics can be assured that Jaki never implied that Christianity *created* science any more than he would have argued that a mother created her own child. Acknowledging the difference between the Creator and His creatures was a central theme in Jaki's arguments. He argued that Christianity provided the nurturing psychology most compatible with the sustained discovery of physical laws and systems of laws. To again quote the Pontifical Academy of Sciences, Jaki argued that the viable birth of science is "intimately tied" because of its "dogma about the divinity of the Incarnate Logos" to "medieval Christianity." Jaki also never denied the contribution of ancient cultures, something a quick perusal of his three hundred and seventy-seven-page book *Science and Creation* will reveal. One could possibly make a distinction between a strong claim that science could *only* have been born of Christianity and a weaker claim that science *was* born of Christianity because the Christian worldview aided the breakthrough. Both claims are valuable in looking to the future of science.

His is not an easy thesis to grasp, to be sure, but then again, neither is the process of gestation, birth, and sustained life to which he compared the development of science. To grasp this complexity is to grasp why Jaki also compared the salvation of science to the Saving Birth of Christ, the central point of salvation history. Jaki's thought requires a broader thinking than a single discipline can provide; it requires a systematic approach where one puts the story together piece by piece in a logical order, but never forgets how each piece fits into the whole picture, a picture that includes science, society, religion, history, and maybe more. Complex as it may be to assimilate, once understood, it is eloquently simple. For now, the oscillating of both the definition of science and the birth of science will probably continue, which is not at all a negative thing. Controversy, if it is ordered, is necessary for ideas to find their place. This book is intended to be used as a reference, which is why the chapters and sections form an outline for easy access. If a reader gains anything from this book, let it be that one may confidently say that *Catholic dogma positively and directly influenced the Scientific Revolution.*

I encourage you, in closing, to reflect on the implications of this knowledge. It should instill in young Catholics courage to pursue scientific fields of study with the assurance that Catholics have a legitimate and significant place in the development of science. It should assure any Catholic that the assertion that science and religion cannot ultimately conflict with one another is an assertion that infuses more freedom in intellectual pursuits than restriction. Indeed to do science well, one may say that a working knowledge of Catholic dogma is beneficial. To know what directly contradicts the dogmas of revealed religion and to make such distinctions offers guidance to the scientist, as was shown by the accomplishments of the medieval Catholic scholars. Jaki's claim about the birth of science demonstrates the axiom, "Truth cannot contradict truth."

Consider how the clarity given to us by Fr. Jaki allows a more appreciative approach to evolution. There is much to be appreciated, studied, and developed in the exact science of it. There are possible cures to be found by studying how genetic mutations affect populations. There are possible benefits to animal populations to be found by understanding how those populations evolve. There are lifetimes of work to be done in studying the exact science of evolution. The contradictions in evolutionary theory arise in the grafting of a materialistic ideology onto that science, which both a Catholic and a scientist should avoid.

Consider the implications in neuroscience. So much of the findings are taken as evidence for materialism, when in fact they can never be such. That does not mean that the brain is not a significant influence on the mind, and there is no contradiction with Catholic dogma to study the brain so long as the soul is not denied. Perhaps if it were better understood that neuroscience belongs to the realm of the physical, and psychology belongs to the realm of both body and soul, then better advances could be made in medicine and treatment that heal the whole person.

Consider the implications of Jaki's work in the fields of physics and cosmology. Perhaps the multiverse theory is not a complete contradiction of Catholic dogma. To the extent that it hypothesizes the possibility of the existence of other worlds, it is consistent with the Condemnations of 1277. Proposition 27 asserted that God can make as many worlds as He wills by rejecting the statement, "That the first cause cannot make more than one world." These are the reasons I find Jaki's work exciting. By providing clarity, his conclusions offer direction and encouragement for the student of science, not just Catholics, but all students.

Should my work in this matter be continued, I, a Catholic convert from pagan feminism, a mother of thirteen children—seven of whom were born and six of whom died in the womb—and a former industrial senior research scientist with a doctorate in chemistry specializing in nanometer-scale materials, propose a further development to Jaki's thesis. To achieve a more ecumenical approach, perhaps the theological history of science can be presented with a Marian character, which is a reasonable extension of the "birth" and "stillbirth" analogy already. How might this be done? I think it begins with education. It begins be instructing students in the exact, hard, and quantitative sciences without overlaying ideology onto them from either side, religious or atheist. Physics, chemistry, and even biology should be taught strictly as hard sciences. A believer, just as the Biblical people, the early Christians, the scholars of the Middle Ages, and Christians today, will see the exact science as the handiwork of God. The non-believer may not, but then again, he may be awed by the order and discover the design in it for himself. Science, exact science that is, ought to be a place believer and non-believer alike can come together as thorough materialists. After all, a believer and a non-believer could cook a meal together and sit down to enjoy it together too, even if the believer thanks God for the gift of food before eating.

In more theological terms, this approach could be developed by considering 1) how the Holy Church is a Mother herself, guiding and encouraging, vetoing what is wrong for her children and proposing a better way for the human family through friendship; and how 2) Christians even of the first century said that the "world was created for the sake of the Church."[2] The past is past. What counts in the study of history is where the instruction leads the next generations, and Modern Science—mature and independent as this child of the human intellect may seem—is desperately in need of its Mother.

---

[2] Jeff McLeod, "The Universe Was Created for the Sake of the Church," *Catholic Stand* (August 27, 2013) at http://catholicstand.com/the-universe-was-created-for-the-sake-of-the-church/, quoting the *Catechism of the Catholic Church* §760.

Bibliography

Aaboe, A. "Scientific Astronomy in Antiquity." *Philosophical Transactions of the Royal Society of London* A. 276, no. Series A, Mathematical and Physical Sciences (1974): 21-24.

Adams, James. *The Republic of Plato: Edited with Critical Notes, Commentary, and Appendices.* Second. Vol. II. Cambridge: Cambridge University Press, 1963.

Aertsen, Jan A., Kent Emery, Andreas Speer, and Walter de Gruyter. *Nach der Verurteilung von 1277 / After the Condemnation of 1277: Philosophie und Theologie an der Universität von Paris im letzten Viertel des 13. Jahrhunderts. Studien und Texte / Philosophy and Theology at the University of Paris in the Last Quarter of the Thirteenth Century. Studies and Texts.* Berlin, New York: Walter de Gruyter GmbH& Co., 2000.

Agnihotri, V. K. *Indian History: With Objective Questions and Historical Maps.* India: Allied Publishers Private Limited, 2007.

Albiruni, Athar-ul-Bakiya of. *The Chronology of Ancient Nations: An English Version of the Arabic Text "Vestiges of the Past."* Translated by D. Edward Sachau. London: W. H. Allen & Co., 1879.

Alexandria, Clement of. *The Exhortation to the Greeks, The Riches of Man's Salvation, and the Fragment of an Address Entitled To the Newly Baptized.* Translated by G. W. Butterworth. London: William Heinemann, 1919.

Alfred North Whitehead – Biography. The European Graduate
    School. http://www.egs.edu/library/alfred-north-
    whitehead/biography/ (accessed October 30, 2013).

Anonymous. *Leucippus (5th C. BCE).* n.d.
    http://www.iep.utm.edu/leucippu/ (accessed
    October 30, 2013).

Antonio Spadaro, S.J. "A Big Heart Open to God." *America
    Magazine,* September 30, 2013.

Aquinas, Thomas. *Summa Theologica, Latin-English Edition, vol.
    1.* NovAntiqua, n.d.

——. "De Motu Cordis, Translated by Gregory Froelich." *The
    Dominican House of Studies.* n.d.
    http://dhspriory.org/thomas/DeMotuCordis.htm
    (accessed November 10, 2013).

——. *Summa Contra Gentiles.* Translated by Anton C. Pegis. n.d.

Arendzen, John. *Manchianism.* The Catholic Encyclopedia,
    New Advent. 1910.
    http://www.newadvent.org/cathen/09591a.htm
    (accessed October 20, 2013).

Aristotle. "Categories." n.d.

——. *Metaphysics.* n.d.

——. *On the Heavens.* n.d.

Augustine. *A Select Library of the Nicene and Post-Nicene Fathers of
    the Christian Church, St. Augustine, The City of God,
    Volume II.* Edited by Philop Schaff. Grand Rapids:
    William. B. Eerdmans Publishing Co., 1956.

—. *Sancti Aureli Augustini De Genesi ad litteram libri duodecim, in Corpus Scriptorum Ecclesiasticorum Latinoram.* Edited by J. Zycha. Vol. Volume XXVIII. Vienna: F. Tempsky, 1894.

Bacon, Roger. *Medieval Sourcebook: Roger Bacon: On Experimental Science 1268, The Early Medieval World, 369-376.* Fordham University. Edited by Oliver J. Thatcher. Vol. V. Milwaukee: University Research Extension Co., 1901.

Barr, Stephen M. "Science Seeking Understanding." *First Things,* June/July 2009.

Bath, Adelard of. *Conversations with His Nephew: On the Same and the Different, Questions on Natural Science and On Birds.* Edited by Charles Burnett. Translated by Charles Burnett. Cambridge: Cambridge University Press, 1998.

Berryman, Sylvia. *Democritus.* Edited by Edward N. Zalta (ed.). n.d. http://plato.stanford.edu/archives/fall2010/entries/democritus/ (accessed November 9, 2013).

—. *Leucippus* . Edited by Edward N. Zalta. n.d. http://plato.stanford.edu/archives/fall2010/entries/leucippus/ (accessed November 9, 2013).

Boylan, Michael. *Hippocrates.* 2002/2005. http://www.iep.utm.edu/hippocra/ (accessed October 18, 2013).

Brabant, Siger of. "The Eternity of the World." Edited by translated by Peter King at. n.d. http://individual.utoronto.ca/pking/translations/SIGER.Eternity_of_World.pdf.

Buridan, Jean. *Quaestiones super quattuor libris de cælo et mundo.* Edited by E. A. Moody. Cambridge: The Medieval Academy of America, 1942.

Butterfield, Herbert. *The Origins of Modern Science.* New York: The Free Press, A Division of Simon & Schuster Inc., 1957.

Church History Study Helps: The Alexandrian School. n.d. http://www.theologywebsite.com/history/alexandria. shtml (accessed November 11, 2013).

Clark, R. T. Rundle. *Myths and Symbol in Ancient Egypt.* London: Thames & Hudson, 1959.

Cochrane, Charles N. *Christianity and Classical Culture: A Study of Thought and Action from Augustus to Augustine.* Oxford: Oxford University Press, 1944.

Collingwood, Robin George. *The Idea of Nature.* Oxford: Oxford University Press, 1945.

Couprie, Dirk L. *Anaximander.* 2001/2005. http://www.iep.utm.edu/anaximan/ (accessed October 17, 2013).

—. "How Thales Was Able to 'Predict' a Solar Eclipse Without the Help of Alleged mesopotamian Wisdom." *Early Science and Medicine* 9, No. 4 (2004): 321-337.

Cowell, E. B., ed. *Maitri Upanishad, Sanskrit Text with English Translation.* Translated by E. B. Cowell. London: Asiatic Society of Bengal, 1870.

Crawford, Harriet. *Sumer and the Sumerians.* Second. Cambridge: Cambridge University Press, 2004.

Crombie, Alistair Cameron. *Robert Grosseteste and the Origins of Experimental Science 1100-1700.* Oxford: Clarendon Press, 1953.

—. *The History of Science from Augustine to Galileo.* Mineola: Dover Publications, Inc., 1995 reprint.

Danielou, Jean Marie, S. J. *The Bible and the Liturgy.* Notre Dame: University of Notre Dame Press, 2002.

Dawson, Christopher. *Progress and Religion.* Garden City: Doubleday, 1960.

Deussen, Paul. *The Philosophy of the Upanishads.* Translated by M. A. Rev. A. S. Geden. Edinburgh: T. & T. Clark, 1906, 1908.

Duhem, Pierre, *Essays in the History and Philosophy of Science.* Edited by Roger Ariew and Peter Barker. Translated by Roger Ariew and Peter Barker. Indianapolis: Hackett Publishing Company, 1996.

—. *Etudes sur Leonard de Vinci.* Vol. I. Paris: Hermann Publishers, 1906.

Eddington, Sir Arthur. *The Nature of the Physical World.* London: Cambridge University Press, 1928.

Ferguson, Everett. *Backgrounds of Early Christianity.* Grand Rapids: William. B. Eerdmans Publishing Co., 1987, 1993, 2003.

Fung, Yu-Lan. "Why China Has No Science: An Interpretation of the History and Consequences of

Chinese Philosophy." *The International Journal of Ethics*, 1922: 237-263.

Gandhi, M. K. *A Dialogue between an Editor and a Reader, Hind Swaraj, or Indian Home Rule.* 1938. http://www.mkgandhi.org/swarajya/ (accessed October 10, 2013).

Gooch, Jason. "The Effects of the Condemnation of 1277." *The Hilltop Review* 2 (2006): 34.

Grant, Edward. *A Source Book in Medieval Science.* Cambridge: Harvard University Press, 1974.

—. *The Nature of Natural Philosophy in the Late Middle Ages.* Washington, D.C.: The Catholic University of America, 2010.

Haffner, Paul. *Creation and Scientific Creativity: A Study in the Thought of S. L. Jaki.* 2nd. Herefordshire: Gracewing, 2009.

Hannam, James. *The Genesis of Science: How the Christian Middle Ages Launched the Scientific Revolution.* Washington, D.C.: Regenery Publishing, Inc., 2011.

Harris, Sam. *Free Will.* New York: Free Press, 2012.

—. *The Moral Landscape: How Science Can Determine Human Values.* New York: Free Press, 2010.

Heller, Michael. *Ultimate Explanations of the Universe.* Poland: TAiWPN Universitas, 2009.

Herodotus. *The Histories, with an English Translation.* Translated by A. D. Godley. Cambridge: Cambridge University Press, 1920.

Hora, S. L. "History of Science and Technology in India and South-East Asia." *Nature*, 1951: 64-65.

Huffman, Carl. *Pythagoras*. Edited by Edward N. Zalta. Fall 2011. http://plato.stanford.edu/archives/fall2011/entries/pythagoras/ (accessed October 18, 2013).

Hyman, Athur, James J. Walsh, and Thomas Williams, ed. *Philosophy in the Middle Ages: The Christian, Islamic, and Jewish Tradition*. Third. Indianapolis: Hackett Publishing Company, Inc., 2010.

Jaki, Stanley L. *A Late Awakening and Other Essays*. Port Huron, MI: Real View Books, 2004.

—. *A Mind's Matter: An Intellectual Autobiography*. Grand Rapids, MI: William. B. Eerdmans Publishing Co., 2002.

—. *Means to a Message: A Treatise on Truth*. Grand Rapids, MI: William. B. Eerdmans Publishing Company, 1999.

—. *Numbers Decide and Other Essays*. Pinckney, MI: Real View Books, 2003.

—. *Questions on Science and Religion*. Pinckney, MI: Real View Books, 2004.

—. *Science and Creation: From Eternal Cycles to an Oscillating Universe*. Edinburgh: Scottish Academic Press, 1986.

—. *Science and Religion: A Primer*. Port Huron, MI: Real View Books, 2004.

—. *The Absolute Beneath the Relative and other Essays.* Lantham: University of America Press, 1988.

—. *The Drama of Quantities.* Port Huron, MI: Real View Books, 2005.

—. *The Relevance of Physics.* Chicago: University of Chicago Press, 1966.

—. *The Road of Science and the Ways to God: The Gifford Lectures 1975 and 1976.* Chicago, Edinburgh: University of Chicago Press, Scottish Academic Press, 1978.

—. *The Savior of Science.* Grand Rapids, MI: William. B. Eerdmans Publishing Company, 2000.

Kantilya. "Indian History Sourcebook: The Arthasastra." *Fordham University.* c. 250 B.C. http://www.fordham.edu/halsall/india/kautilya1.asp (accessed October 1, 2013).

Kendra, D. P. Agrawal and Lok Vigyan. "The Needham Question: Some Answers." *Indian Science*, n.d.

Koyre, Alexander. *From the Closed World to the Infinite Universe.* US: JHU Press, 1957.

Legge, J., trans. *Texts of Taoism.* New York: Julian Press, 1959.

Lendering, Jona. *The First Circumnavigation of Africa.* n.d. http://www.moellerhaus.com/Persian/Hist01.html (accessed November 19, 2013).

Lin, Justin Yifu. "The Needham Puzzle: Why the Industrial Revolution Did Not Originate in China." *Economic Development and Cultural Change*, Chicago: University of Chicago Press, 1995.

Lindberg, David C. *The Beginnings of Western Science: The European Scientific Tradition in Philosophical, Religious, and Institutional Context, Prehistory to A.D. 1450.* Chicago: The University of Chicago Press, 1992.

Magnus, Albertus. "About Fate." *Corpus Thomisticum.* n.d. http://www.corpusthomisticum.org/xpz.html (accessed November 10, 2013).

McEvoy, James. *Robert Grosseteste.* Oxford: Oxford University Press, 2000.

McHenry, Robert. *Thales of Miletus: The First Scientist, the First Philosopher.* Inc. Encyclopedia Britannica. n.d. http://www.britannica.com/blogs/2010/04/thales-of-miletus-hero/ (accessed October 15, 2013).

McLeod, Jeff. "The Universe Was Created for the Sake of the Church." *Catholic Stand*, August 27, 2013.

Morris, Christopher G. *Academic Press Dictionary of Science & Technology.* San Diego: Academic Press, 1992.

Narasimha, Roddam. "The Indian Half of Needham's Question: Some Thoughts on Axioms, Models, Algorithms, and Computational Positivism." *Interdisciplinary Science Reviews* 28, No. 1 (2003): 1-13.

Needham, Joseph. *Science and Civilisation in China.* Vol. 2: History of Scientific Thought. Cambridge: Cambridge University Press, 1956.

—. "Science and the Crossroads." *2nd International Congress of the History of Science and Technology.* London: International Congress of the History of Science and Technology, 1931. Foreward.

Neugebauer, O. "The Origin of the Egyptian Calendar." *Journal of Near Eastern Studies*, 1942: 396.

Newton, Isaac. *Principia, A New Translation.* Translated by I. B. Cohen and A. Whitman. Berkeley: University of California Press, 1999.

Numbers, Ronald L. *Galileo Goes to Jail and Other Myths About Science and Religion.* Cambridge and London: Harvard University Press, 2009.

O'Connor, Bernard. "The Catholic Church and Galileo." EWTN. 2009. http://www.ewtn.com/library/ISSUES/churchgalile o.htm (accessed November 1, 2013).

Origen. *Contra Celsum.* Translated by Henry Chadwick. Cambridge: Cambridge University Press, 1953.

Plato. *Republic in Plato: The Complete Works.* Translated by Benjamin Jowett. Latus ePublishing, 2012.

Plinio, Prioresch. *A History of Medicine: Roman Medicine.* Omaha: Horatius Press, 1998.

Pope St. John Paul II. *Fides et Ratio.* Vatican City: Vatican, 1998.

Roberts, Alexander, Sir James Donaldson, and Arthur Cleveland Coxe. *Ante-Nicene Fathers Volume I: The Apostolic Fathers, Justin Martyr, Irenaeus.* New York: Charles Scribner's Sons, 1925.

—. *Ante-Nicene Fathers Volume II: Fathers of the Second Century: Hermes, Tatian, Athenagoras, Theophilus, and Clement of Alexandria.* New York: Charles Scribner's Sons, 1925.

Rural Roads: A Lifeline for Villages in India. n.d.
http://web.worldbank.org/ (accessed October 4,
2013).

Sample, Ian. "What is the thing we call science? Here's one
definition ..." March 3, 2009.
http://www.theguardian.com/science/blog/2009/m
ar/03/science-definition-council-francis-bacon
(accessed September 15, 2013).

Shields, Christopher. *Aristotle.* Edited by Edward N. Zalta.
2013.
http://plato.stanford.edu/archives/fall2013/entries/
aristotle/ (accessed October 18, 2013).

Smith, Vincent Edward. *The General Science of Nature.*
Milwaukee: Brice, 1958.

Sorondo, Nicola Cabibbo and Marcelo Sánchez. "Pontifical
Academy of Sciences, announcement of the death of
Prof. Stanley Jaki." April 8, 2009.
http://www.sljaki.com/Pontifical_Academy_en.html
(accessed November 2, 2013).

Sparavigna, Amelia Carolina. "Reflection and refraction in
Robert Grosseteste's De Lineis, Angulis et Figuris."
Cornell University. n.d.
http://arxiv.org/ftp/arxiv/papers/1302/1302.1885.p
df (accessed October 20, 2013).

Taisbak, Christian Marinus. *Euclid.* 2013.
http://www.britannica.com/EBchecked/topic/1948
80/Euclid.

The Knox Translation Bible. Baronius Press Ltd., 2013.

Thorndike, Lynn. *A History of Magic and Experimental Science.* Vol. 11. 14 vols. Colombia: Colombia University Press, 1923.

Thorndike, Lynn. "The True Roger Bacon." *The American Historical Review* (The Macmillan Company) XXI (1916).

Tkacz, Michael W. "Aquinas vs. Intelligent Design." *Catholic Answers Magazine* 19, no. 9 (November 2008).

Toulmin, Stephen. *Foresight and Understanding: An Enquiry into the Aims of Science.* New York: Harper & Row, 1961.
Turner, William. *William of Auvergne.* The Catholic Encyclopedia, New Advent. 1912. http://www.newadvent.org/cathen/15631c.htm (accessed October 25, 2013).

Weisheipl, James A., O.P., *Albertus Magnus and the Sciences: Commemorative Essays 1980.* Toronto: Pontifical Institute of Medieval Sciences, 1980.

Whitehead, Alfred North. *Science and the Modern World.* Cambridge: The MacMillian Company, 1925.

Wiker, Benjamin. *The Catholic Church and Science: Answering the Questions Exposing the Myths.* Charlotte: TAN Books, 2011.

Williams, Henry Smith and Edward Huntington. *A History of Science.* Vol. 1, Book I. New York: Harper & Brothers, 1904.

Woods, Thomas E. *How the Catholic Church Built Western Civilization.* Washington D.C.: Regnery Publishing, Inc., 2005.

About the Author

STACY A. TRASANCOS is a professor of chemistry at Holy Apostles College and Seminary.

Her website: stacytrasancos.com

CPSIA information can be obtained
at www.ICGtesting.com
Printed in the USA
BVHW070305150120
569367BV00002B/213/P